Shepherd's Purse

Organic Pest Control Handbook

FOR HOME AND GARDEN

REVISED EDITION

BY PEST PUBLICATIONS

THE BOOK PUBLISHING COMPANY
SUMMERTOWN, TENNESSEE 38483

The authors wish to thank the many people involved in the production of this book. Some contributed directly and others indirectly. To each of them our heart-felt thanks.

We would also like to thank a certain lovely dragonfly that kept us company one day on the steps of our house next to the garden while writing parts of this book. Every now and then she would take off and do a loopty-loop in the air with her noisy wings and land right back beside us for a while then fly off again to catch another mosquito or leafhopper. We finally had to abandon our sunny steps because we kept putting down our pens and pencils and watching intently the marvels of our garden and our world.

Author's Note: Special thanks to the following people—Dorothy Bates, W.O. Wilson, Eleanor Dale Evans, Bob and Cynthia Holzapfel, and Richard Martin.

Shepherd's Purse is the name of an insect-eating plant similar to *Drosera* (sundew plants), *Sarracenia* (pitcher plants) and the Venus fly trap.

Design and Layout: Eleanor Dale Evans
Illustrations: Richard Martin

Published by Book Publishing Company, Summertown, TN 38483

ISBN 0-913990-98-1

Shepherd's purse : organic pest control handbook / by Pest Publications — Rev. ed.
p. cm.
Includes bibliographical references (p.) and index.
ISBN 0-913990-98-1 : $9.95
1. Insect pests—Biological control—Handbooks, manuals, etc. 2. Insect pests—Control—Handbooks, manuals, etc. 3. Organic gardening. I. Pest Publications (Firm)
SB933.3.S48 1992
635'.0497--dc20 92-25525
CIP

Printed in MEXICO

Every effort has been made to present safe controls, but the use of them is at the reader's risk. Reference to commercial products and their trade names is made for purposes of clarity and no discrimination of other companies with similar products is implied.

CONTENTS

Introduction

Most gardeners encounter an insect problem during the growing season. Insects, good and bad, are the most common "wild animals" we observe in the garden. The question most gardeners ask is "What do I do about the bugs?" There is no absolute answer to that question, but we plan to outline some alternatives to the use of persistent or broad range poisons. This outline should help you with your own experimentation, the trial and error process that is a large part of the fun and delight in gardening.

Some preventive measures, cultural and biological, are necessary for insect management. We have found that doing nothing or very little else in the way of insect control turns out to be the safest and most effective insect management. Every now and then pest populations can get out of balance and we have to do something about the bugs, but that happens far less than we ever could have imagined when we first began gardening.

Most beginning gardeners anticipate that every plant will be eaten down to the ground if something is not done about the bugs. After the first season we realized how foolish our anxieties really were and it was only our lack of experience and knowledge that contributed to that fear.

Biological Control

For gardening purposes the word "control" means maintaining pest populations within a tolerable range. What you want to tolerate is up to you. Usually control does not mean elimination or eradication. Eradication can be impossible, expensive and dangerous.

Biological control simply implies using living organisms or their products to control pests. It has successfully controlled pest plants (weeds), pest insects and even pest rabbits (vertebrates).

Biological techniques may involve simple things such as:

1. Handpicking
2. Use of natural predators (organisms that eat pests whole)
3. Parasites (organisms that eat pests from the inside)
4. Microbials (microscopic organisms that make pests sick)
5. Companion planting, where one plant protects another
6. Natural scents (pheromones) and tastes to repel, trap or confuse insects

Organic Insect Control

Organic control involves all the biological control methods listed above with three additional considerations:

1. Soil health and fertility (composting, etc.)
2. Knowledge of insect biology to allow for cultural controls like crop rotation
3. Trapping — using simple physical controls

Integrated Pest Management

Integrated pest management, or IPM, is similar to the concepts above and is useful for both commercial situations and home, yard and garden settings. On occasion IPM uses synthetic pest poisons, if necessary. Some IPM concepts useful for gardeners to keep in mind are:

1. Identify pest insects, understand their biology and natural enemies.
2. Monitor insect population densities either directly or with traps.
3. Determine level of plant injury related to population densities.
4. If population is above a threshold of tolerance, use a combination of control strategies including natural enemies (biological control), cultural controls and selective pesticides.
5. Evaluate the effect of the strategy by monitoring. (Back to step 2)

IPM aims to work with natural balances as much as possible. Commercial farmers like its cost effectiveness and how it often avoids insect resistance and environmental problems. Monitoring is the heart of IPM.

A Word About Plant Diseases

Many plant diseases are spread by insects or caused by soil fertility imbalance. House plants that become diseased are usually attacked by whitefly and aphids. These inconspicuous insects suck plant juices by piercing plant tissues. They transmit viruses that they inject into plants with their saliva. Further, the wounds they make start to decay and are susceptible to airborne diseases that land on the wounds. Spraying the plants with soap or releasing *Encarsia Formosa* (white fly parasite) in the house will prevent plant injury.

Aphids also excrete a sweet juice called honeydew. Honeydew can become a breeding place for molds. Sometimes removing the honeydew with a cotton swab will prevent diseases like sooty mold. However, the root cure for many diseases is removing the insect spreading the disease.

Diseased outdoor plants may indicate that soil fertility has declined beyond the point of supporting the plants. However, even with good fertility, old plants become susceptible to disease. Citrus orchards are replanted after a number of years for this reason. The best remedy is to replant with young plants.

The gardener can spread disease in the garden by handling wet plants or by touching a diseased leaf and then handling other plants. Cut off and destroy diseased leaves and wash your hands before touching another plant. Some disease organisms live in the soil. Compost heats up considerably as it forms and is free of most soil borne diseases.

Insect control and good soil enriched by compost are the keys to avoiding most disease problems. Crop rotation also prevents plant diseases in the garden.

Biological Insect Control

Augmenting Beneficial Organisms

Unknown to most beginning gardeners, there are many beneficial parasites and predators that are natural enemies of pests. Sometimes natural controls are already present in your garden, but are either in very small numbers or their build-up is delayed. Here the gardener does well to augment the beneficial organisms. Increasing their number may involve early or concentrated release of them.

Beneficial insects, either naturally occurring or commercially bought, often require nectar, pollen and water. Blooming flowers provide nectar and pollen. We suggest that your garden contain flowers that are varied enough in type to be in bloom as much of the year as possible. For a water source to increase the humidity in the garden, we recommend digging two or three holes about 10 inches in diameter and 8 inches deep, fit with clay flower pots and fill with water. If mosquitoes are a problem, one small briquet of Bti (page 9) or Bti spray in each hole or flower pot will control mosquitoes. Successive wetting and drying will not affect the potency of Bti, but it will dissolve over time and need to be replaced.

Toads eat many insects. You can encourage them to work in your garden by placing an inverted clay flower pot with a large hole chipped out of the rim placed in the shade near a water hole to provide a nice home.

Braconids are another example of nature helping the gardener. They are small wasps that sting large pest caterpillars like tomato hornworms. As they sting, they lay eggs into the caterpillar. The eggs hatch and braconid larvae develop in the living caterpillar. Finally the larvae eat through the caterpillar's skin, usually on its back. Here they immediately spin cocoons that resemble small rice grains, The caterpillar is still alive, but it will soon die, leaving the adult wasps to hatch and attack other caterpillars. We don't have to buy this naturally occuring insect.

Unfortunately these little fellows and other helpful organisms are often more sensitive than the pests to pesticides, or broad spectrum poisons. These poisons kill a wide range of insects and living things. Since there are fewer numbers of beneficial parasites and predators, poisons destroy the delicate balance of nature between helpful and harmful insects. Carbaryl (*Sevin*) and malathion are examples of broad spectrum petroleum based poisons often recommended by some county agents and dispensed by discount stores.

We also caution against the indiscriminate use of broad spectrum botanical poisons like rotenone and pyrethrin with piperonyl butoxide. These botanical poisons can kill most any insect including beneficial ones. They do break down quickly though, and don't persist in the environment like some synthetic poisons. Still, care should be exercised. We recommend them for occasional spot treatments of acute problems, but not for spraying every plant. Rotenone is a poison.

Commercially Available Organisms

Commercially available beneficial insects are listed for each pest insect in this book. The following is a discussion of three beneficial insects that should be used for general control of many insects in the yard and garden. For home gardens we suggest using the smallest amount available, applying three releases at two week intervals in the early spring when flowers are blooming after the last frost. An early fall release may be necessary in areas with a long growing season.

Trichogramma are insects no bigger than the period at the end of this sentence. They sting pest eggs and lay their own eggs inside them. Trichogrammid larvae hatch and kill the pest eggs. They are sold commercially on cards of about 400 parasitized eggs. When they arrive, the trichogramma adults are just about to emerge from the eggs and begin their life cycle. These minute beneficial insects help protect Texas cotton and Washington apples from different caterpillar pests.

A trichogramma release stand can be made in gardens from a one foot stake driven a few inches into the ground. Halfway up the stake apply a two inch band of petroleum jelly all around to keep ants away from the eggs. Staple or tack the card with trichogramma to the side of the stake near the top. Then as a final measure to protect the card from rain, staple or tack a three inch cardboard "roof" to the top of the stake. Most trichogramma hatch and fly away within three days, but leave the card undisturbed for a week to allow for cooler weather or slowpokes. Lightly water the garden for about two weeks after the release to ensure the first generation's survival. Hot dry weather causes the greatest loss of trichogramma. Warm humid weather with rain provides ideal conditions for their growth and production of succeeding generations.

During winters, trichogramma populations decrease drastically, so you need to release them every spring. Their chances of survival over winter is uncertain and even if they do survive, there may not be enough of them to help you unless you supply more.

Lacewings are net-winged, pale green insects that flutter like butterflies. They are available from several companies; check the list of suppliers in the appendix. Their alligator-shaped larvae are voracious predators of pest insects.

Ladybug (ladybird) beetles and their larvae are also valuable predators of aphids and other pests and pest eggs. To encourage them to remain in the garden you can provide some food for them, called "wheast". You can buy wheast when you buy ladybugs or ask for a recipe. Delaying application of ladybugs until garden plants are larger and flowers are blooming is an alternative to providing wheast for them.

Praying mantis egg masses can be bought and set in your garden. They hatch into minute insects that grow to be effective, beneficial predators. During walks in the woods in the fall you may spot egg masses that you can clip off for free. Clip off the whole mass on its twig. Don't bring them inside or the heat may kill them or hatch them too soon. Place them in a protected bush outside for the winter near you garden. Goldenrod is a favored habitat of the praying mantis.

Control with Micro-Organisms (Microbials)

Insects have their own diseases much like plants do. The insect disease organisms are naturally present in your garden already, but at too low a level to balance pest populations. To achieve a balance, increase them. This is microbial control. Several microbial control agents are commercially available to help you. They are very selective and are harmless to good insects. This is important, since you don't want to disturb good insects the way broad spectrum poisons do. The following table lists some available microbial controls. (Mention of Ringer products is for illustration only. Many other manufacturers produce similar products.)

Commercially Available Microbial Controls

Microbial Control	Product	Pest	Characteristics
Bacillus thuringiensis *(Bt)*	Ringer Vegetable Insect Attack or Ringer Caterpillar Attack	only caterpillars	sprayed or dusted, must be eaten by the insect
Bacillus thuringiensis israelensis *(Bti)* a special strain of Bt	Ringer Mosquito Attack	only mosquito and black fly larvae	kills larvae in water, sprayed or applied as a floating ring
Bacillus popilliae (milky spore disease)	Ringer Grub Attack	grubs of Japanese beetles	applied to soil, very long lasting
Insect-eating nematodes *(Nc)*	Scanmask, others	soil pests, boring pests	seeks out pests, several ways to apply
Nosema locustae	Ringer Grasshopper Attack	grasshoppers and some types of crickets	applied as bait, must be eaten by the insect
Codling Moth Granulosis virus (CMGV)	Decyd ™	caterpillars of codling moth	must be eaten by the insect
Nuclear-polyhedrosis virus (NPV)	Gypchek ™	gypsy moth caterpillars	must be eaten by the insect
Bt (sd)	Ringer Colorado Potato Beetle Attack	Colorado potato beetle, elmleaf beetle	must be eaten by the insect

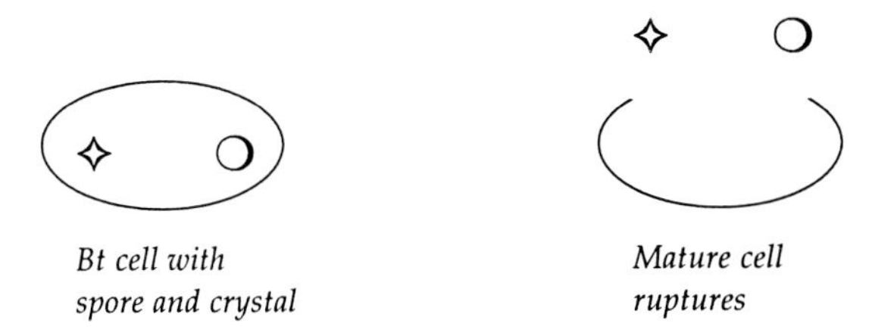

Bt cell with spore and crystal

Mature cell ruptures

Bacillus Thuringiensis or "Bt" is a rod shaped bacterium. Bt is sold as a mixture of the resting spore stage and diamond-shaped protein crystals that form near the spores. It affects only caterpillars that eat it. The crystals stop the insects' feeding, then the spores germinate and the insects die. Packages of Bt can be stable for as many as ten to fifteen years, but is inactivated by direct sunlight after a couple of weeks. It is sold as *Thuricide, Dipel, Caterpillar Attack, Ringer Vegetable Insect Attack* and others.

Bti (*Bacillus thuringiensis israelensis*) is a special strain of Bt active only against mosquito wrigglers and black fly larvae in water. It too must be eaten to kill them. It is safe for all beneficial aquatic insects like mayflies. One can either spray Bti or toss floating rings of it onto water. It is sold as *Mosquito Attack, Vectobac,* etc.

Bacillus Popilliae is a soil inhabiting bacterium. It infects only grubs of the Japanese beetle and a few species of June beetles. It turns the normally clear blood of these grubs a milky color, so it's also known as "milky spore disease". All the grubs feed on roots and are pests. However, it does not kill several species of pest grubs, so precise grub identification by your county agent or lawn and garden center is necessary to be sure you have a grub it will work on. Since Japanese beetle grubs are the most common ones, some people just treat with milky spore as a preventative.

Milky spore is sold under trade names *Doom* and *Ringer Grub Attack*. Apply it to grassy areas and around orchards where adult beetles lay eggs and grubs develop. After application, allow one to two years for it to spread through the soil. One application of milky spore can last up to 20 years. Both grubs and adults damage plants so control the beetles by controlling the grubs.

Insect-eating nematodes (Nc; *Neoaplectana carpocapsae,* alternately named *Steinernematoda carpocapsae*) are beneficial, microscopic, threadlike relatives of the worm family which attack and parasitize many pest insects. They were first discovered in America in codling moth larvae in Beltsville, Maryland, but since then have been found in almost every geographical area in low concentrations. Over 230 different pests are susceptible to them, including squash vine borers and cucumber beetles. Nc nematodes actively seek out the insects they prey upon. They will crawl into borer holes if they are sprayed or injected very close to their burrows.

Just as there are good and bad insects with different life styles, so there are different kinds of nematodes. Most nematodes are free-living and are present in all soil where they help to break down solid organic material, such as in compost. A few other nematodes attack some plant roots. Nc nematodes are different; they only attack pest insects and feed on them where there is a moist, dark environment such as in soil.

Nematodes enter the insect they attack either through its mouth or the spiracles, the breathing tube passages. They enter the blood-filled body cavityand begin feeding. Their method of entry into the cavity is by sheer mechanical force, not a chemical reaction. While feeding, they release a bacterium in their intestines which is emitted in the fecal material they expel. The bacterium "mumifies" the insect or preserves it while the nematodes feed and reproduce inside. The insect usually dies within 24 hours after the nematodes have entered its mouth. Ten to fourteen days later nematodes emerge through the host insect's body wall and search for new insects to infect. During this stage their digestive system has collapsed and they do not feed. Their entire body is enclosed in a sheath of cells called a cuticle. They shed this cuticle after they enter the host insect's digestive midgut and before they force their way between the cells of the host insect's intestinal wall. Some nematodes can live in soil for 12 to 15 months without feeding.

Nc nematodes are shipped in a topsoil or peatmoss-like medium for application on the soil. Other producers ship them on a moist sponge. Trade names are *Grub-eater* and *Scanmask* as well as others.

These beneficial nematodes can dry out and die, but a soil that supports plant growth is moist enough for nematodes to survive. When first putting them out though, be careful to have a moist soil and avoid sunlight. Apply at dusk to presoaked soil or wet tree trunks. This helps the nematodes move around.

Nosema locustae is a one-celled protozoan parasite that is mixed into a cereal bait (wheat bran) that kills grasshoppers and crickets. The resting spore stage is sold as *Nolo* and *Grasshopper Attack*. Maximum effect is achieved a year after application.

Codling Moth Granulosis Virus (CMGV) and Nuclear-Polyhedrosis Virus (NPV) are two viruses that effect some pest insects and not others. They are rather potent, but must be eaten to be effective. Similar virus diseases of other pests also exist, explaining why grinding up dead or sick insects and spraying the suspension may help kill other pests. There is a different strain of NPV for each different insect.

Viruses are obligate parasites of living cells. All known biological life can be classified into six major kingdoms.
1. Bacteria
2. Viruses
3. Protozoa
4. Fungi
5. Plant
6. Animal

Companion Planting

Companion planting or repellent planting means mixing plants in a row so that plants with odors, tastes or other characteristics repellent to pests are planted next to host plants that they like and damage. The insects find the mix of plants not to their liking and leave the garden for a better place to feed. An example is interplanting radishes with cucumbers to repel cucumber beetles. Another is larkspur to repel Japanese beetles. Finding which plants work best for this interplanting is a matter of trial and error in each geographical location.

Mixing flowers in a vegetable garden will help, but may attract some unwanted pests. Further, some plants inhibit others. The following table is a list of further suggestions with which to begin your own experimentation.

Suggested Companion Plantings

Pest	Plant Repellent
Ants	spearmint, tansy, pennyroyal
Bean Leaf Beetle	potato, onion, turnip
Codling Moth	common oleander
Colorado Potato Beetle	green beans, coriander, nasturtium
Cowpea Curculio	garlic, cloves, radish
Cucumber Beetle	radish, tansy
Flea Beetle	garlic, onion, mint
Harlequin Bug	radish, turnips, onion
Imported Cabbage Worm	mint, sage, rosemary, hyssop
Japanese Beetle	garlic, larkspur, red buckeye
Leaf Hopper	geranium
Mexican Bean Beetle	potato, onion, garlic, radish
Root Knot Nematodes	French marigold
Spider Mites	onion, garlic, cloves
Squash Bug	radish, marigold
Squash Vine Borer	cloves, onion, garlic
Stink Bug	radish
Tarnished Plant Bug	garlic, pepper
Thrips	marigold
Tomato Hornworm	marigold, sage
Whitefly	marigold, nasturtium

Traps

Closely related to the ideas of companion planting is the use of products that confuse, lure or trap pests. However, all insect traps are not a good choice even if they are effective. Black light traps kill many valuable beneficial insects like lacewings and trichogramma, but few mosquitoes or other pests. We don't recommend them. Light traps can be useful for monitoring purposes and used with a funnel trap to control some adult stage beetles like June beetles.

The commercial traps available use floral or plant scents or specific sexual scents (pheromones) from specific insects to lure pests. A good example is *Ringer Beetle Trap Attack* for Japanese beetle control. They may also use certain attractive colors, like yellow. Here again, use caution as some of these commercial traps may be too powerful. If you hang up a trap for Japanese beetles, place it away from sensitive plants, like rose bushes, so the beetles will be lured away from there.

You can also make your own trap with "tanglefoot" resins. It can be applied around some trees and bushes to prevent ants and other flightless pests from climbing. Smearing "tanglefoot" on a red plastic apple hung in apple trees will catch the flies that breed apple maggots. (See Glossary for definition of terms)

Cultural Insect Control

Crop rotation helps enrich soil and control insect. It means changing the type of plant at a particular garden location from one year to the next. Generally, three years between planting of vegetables in the same family or class is best. Small gardens may have to be moved after several years in order to achieve good crop rotation.

Good garden sanitation is also important. If you remove diseased or rotting fruit, you will help remove insect hiding places and food.

Another important insect control is fall plowing. Most beginning gardeners forget this task and by doing so create insect problems the following spring. If you plow all old plants and vines under in the fall, many eggs and insects are crushed or dried out. Others are exposed for birds to eat. Some gardeners combine planting a cover crop with fall plowing. After plowing, sprinkle some clover or rye seeds over the soil, rake them in lightly and water. Be sure to use enough seeds to give a thick stand, ¼ to ½ pound of clover seeds/1000 square feet. Plow the clover under the following spring.

Alternate planting methods can also contribute to cultural control. Most people plant their gardens in traditional single rows. An alternative is beds or wide rows. For instance, beds can be two feet wide with a pathway or walkway along the side. There's no need for boards or lumber in making flat beds such as these. This method prevents soil compaction and provides insect control. If vegetables and/or flowers are planted randomly and alternately in beds instead of single file, some insects will be confused, have trouble finding a plant to eat and leave the garden.

Row covers made of cheese cloth netting purchased from a grocery store or cotton based netting closely woven and commercially available, will help deter pests such as Japanese beetles and leaf miners.

Acute Insect Control

A Word for Safety

The word "safe" can be applied to drinking water. All other garden chemicals, homemade and commercial, are questionable. This is especially true of cycloheximide and other seed and leaf fungicides. Consult your medical doctor or poison control center if in doubt. Compressed air sprayers and dust should not be used without a protective mask available at hardware stores. Diatomaceous earth may cause lung silicosis if inhaled and kills some beneficial insects, as well. Always read the label and use eye protection.

Sprays

Insecticidal soaps are special soap solutions to be sprayed on insects. They create a thin film that blocks the tubes that all insects breath through. These fine branching tubes can be blocked by even the thinnest of soap films. However, soap residues on leaf surfaces have no effect on insects or other animals, so the soap spray must be sprayed directly on the insects. Be sure to spray the underside of leaves where many insects are found. When it is recommended for use on caterpillars, it is most effective on smaller ones. Commercial insecticidal soaps are mild and do not burn plants. Some are formulated for flowers, others for vegetables and fruit, dependent on the carriers used. Some commercial insect soaps have as much as 30% alcohol content, even though their label does not state this. Soap spray can adversely affect many beneficials, including dragonflies, so you might use it as spot spraying only.

spraying only.
Homemade soap sprays can also kill insects, interacting chemically with the outer membrane of the insect. Since they can also burn plant leaves, trial and error should be used to determine the brand of soap and concentration you use. We recommend starting with 2-3 tablespoons per gallon of water. Rinse off the plants with water after an hour or two if you are not willing to experiment. This should be sufficient time to affect the insects, but not harm the plants. Soaps with perfume and dyes are not recommended. We've found some liquid dishwashing soaps so mild they can be left on the plants.

Ordinary agricultural lime, used at ¼ to ½ cup per gallon of water, makes another homemade spray that kills some insects and mites and dries out many smaller insects. Its effectiveness depends upon the insect's surface area and weight. Spider mites are tiny "bugs" that are affected by lime. Generally, most insect larvae are sensitive as are others like aphids. Soap and lime mixed together make a potent double-acting spray with the soap helping the lime stick to the insect. Lime also acts as an irritant for adult insects even if it does not kill them. When using lime dust on wet plants, you might want to rinse it off with water after one or two days. Some plants might have an adverse reaction to either soap or lime. Do a trial spraying on a few leaves and wait two or three days to determine the plant's reaction before spraying the whole plant. Vegetables such as cabbage and broccoli like lime. Again, it's a matter of experimentation. We spot spray lime in the garden like we spot spray with soap. Caution: do not overuse lime, because it may change soil PH levels. Compost acts as a good chemical buffer, though, to counteract either too acid or too alkaline soil conditions.

Homemade repellents can be made of garlic, onion, or pepper juices. Insects are not poisoned, but plant surfaces are made repellent either to taste or touch. One recipe uses a garlic clove, one small onion, and ¼ teaspoon cayenne pepper added to one quart water plus one teaspoon liquid dishwashing detergent to make the spray stick to leaves. We found that substituting several hot pepper pods for the cayenne pepper makes a more effective mixture. Finely chop the vegetables or juice them in a blender, then strain the juice through cheese cloth. When working with hot peppers, be careful to protect your hands from the burning juice with rubber gloves.

Handpicking

Handpicking is the oldest known form of insect control. Insects are picked from plants and crushed. Insects like adult potato beetles and tomato hornworms are easily controlled by handpicking. Wear gloves to avoid allergic reactions. Handpicking is actually a biological control, but since it is so fast, we include it under acute controls.

To make the job easier, hose off plants to discourage aphids, spider mites and some other insects who dislike being wet. For another trick, use a small glass jar with ¼ inch of water in the bottom. Hold the jar slightly under a leaf, then flick the bug into the water.

A butterfly net can be made of two coat hangers and a two yard piece of cheese cloth. Capturing one moth in the spring means two to three hundred less cabbage worms to eat your broccoli.

If enough insects are handpicked in spring and early summer, succeeding generations may be severely reduced.

Microbials

Bt and beneficial Nc nematodes are also fast acting biologicals for certain insects and can be considered acute controls.

Creating a Good Garden Environment

We believe that insects attack weak, diseased plants and act as "nature's pruners." A rich soil grows healthy plants which resist insect attack. Each plant requires a unique soil for optimum growth; some require more nitrogen than others. Knowing your plants' nutritional requirements will help you control insects more effectively.

Organic Soil Enrichment

We suggest that your soil be enriched over a period of years using among others:

1. Natural microbial fertilizers, such as *Ringer Restore*
2. Homemade compost, to eventually achieve a 1 foot depth
3. Composted cow manure
4. Cottonseed or soybean meal for nitrogen sources
5. Steamed bone meal or ground rock phosphate for phosphorus sources
6. Granite dust (about ¼ cup per square foot) or green sand, interacting with a high organic matter content in the soil to create a potassium source
7. Green manure and mulches

Most people realize soil in the woods or forest is very rich in humus and good for plants to grow in. There's definitely value in knowing how to make that same woodsy, humus-rich soil in the garden.

Soil contains living micro-organisms, bacteria, fungi, etc., in addition to nonliving particles and humus, decaying organic matter. Chemical fertilizers kill soil micro-organisms, both good and bad, and contribute to nitrate contamination of surface water and water wells. We used commercial fertilizers only when we started gardening, until the soil was enriched with compost, then we discontinued them. A healthy soil is developed over a number of years; it is not an instant process.

Compost is easy for homeowners to make. One recipe calls for equal parts, by weight, of fresh grass clippings, fresh fallen leaves and composted cow manure. Shred the leaves by running over them several times with the lawn mower. In the spring and summer when leaves are not available, substitute hay or wheat straw for leaves (if straw mites irritate your skin after working with straw, apply alcohol on your hands, arms and legs, followed by a warm soapy bath). Experienced compost makers can delete the composted cow manure. Make sure the leaves and grass clippings have not been sprayed with pesticides. Mix in clippings and manure, then moisten and pile in a hill. Fasten plastic garbage bags on the top ⅔ of the hill with stones or bricks to contain odor and prevent the pile from drying too quickly. The hill should be neither too moist nor dry.

Depending upon temperature and weather, compost should be ready after four to six weeks in summer or two to four months in winter. Turn the hill once or twice during that time to mix the outside with the inside. The more you turn the hill the faster the compost is made.

If for some reason the composting and decaying process does not take place, add more composted cow manure or cottonseed or soybean meal. To speed things up, make a volcano-like depression in the center to help get air and moisture to the middle of the hill. A compost inoculent like *Ringer Compost Maker* can increase composting activity and speed. It contains natural soil micro-organisms and sources to increase biological activity.

You can also make compost in plastic garbage bags. Place the ingredients inside a bag and close it with a twist tie. Be sure to perforate the bag with as many small air holes as possible. Turn or shake vigorously 3 or 4 times or more per week to speed up the composting process.

Green manure or cover crops means planting clover, vetch or rye grain seed, then plowing the plants under in the spring.

Weed Control

Weeds are a common gardening problem. Weed seeds may remain viable in the soil for many years. Some can germinate after 25 years. However, seeds buried 2 inches below the surface will not germinate. Some gardeners bury indigenous weed seeds so deeply with compost that they never sprout. Instead of plowing, pull the dying plants up and compost them.

For warmer climates we suggest the following procedure to reduce weed populations. Plow in the early spring. Wait until the weeds start sprouting, then lightly cultivate no more than one inch with a hoe or rake, or spray apple cider vinegar on them. Wait a week or two, then cultivate or spray again. Be careful never to cultivate below an inch or new weed seeds will be brought to the surface, sprout, and grow. Wait another week or two, cultivate or spray, then plant your garden. Of course, watering the garden will force the weeds to germinate faster. Soapy water can be substituted for vinegar to kill very young weed seedlings. Use five tablespoons of liquid dish washing soap per gallon of water. There are differences in soaps, *Ivory* being one that is especially effective, so try several if one doesn't work. The soap or vinegar kills weed leaves which starves the roots. Repeated leaf destruction will eventually destroy the roots entirely.

After your garden comes up, roto-tilling deeply between rows only brings up new weed seeds to the surface. A hand held wick-type applicator filled with soapy water or vinegar can be used between plants and rows after the plants have emerged.

Common Pests of Gardens and Houses

The most troublesome pests tend to be insects, so they predominate this list. Since this is a book for gardeners, most of the pests are plant feeding types that plague gardens, but some house pests are also included. Controls listed may also be used for flowers, shrubs and fruit trees. Many insects feed and reproduce on other plants besides those in the garden.

All pests are presented in alphabetical order with the Latin name following the common name to help you if you'd like to research them further. In some instances many different species of an insect can be found, but the most common variety is listed. The drawings are meant to be representative of the insect form you will most likely encounter.

We describe the pests briefly with biological, cultural and acute controls. Acute controls should only be used when you need drastic help fast, since they may also kill beneficial organisms and upset the balance of nature. The use of long term biological or cultural control for prevention is better.

Insect Anatomy and Life Cycles

Adult insects have a head with two antennae, a thorax (in the middle), and an abdomen, with six legs on the thorax and usually, but not always with a pair or two of wings also on the thorax. Insect eggs usually hatch into larvae, also called caterpillars, grubs, maggots or worms. They transform into the adult form, a fly, butterfly or beetle after going through a pupal or cocoon stage. Other insects have eggs that hatch into nymphs that gradually transform into adult bugs. We describe the life form in detail that usually causes the gardener's direct problems. When lengths are given in inches, these are maximum lengths. The newly hatched insects are usually minute, almost microscopic, but grow rapidly.

Ants *Formicidae family, hymenoptera order*

Fire Ants *Solenopsis geminata*

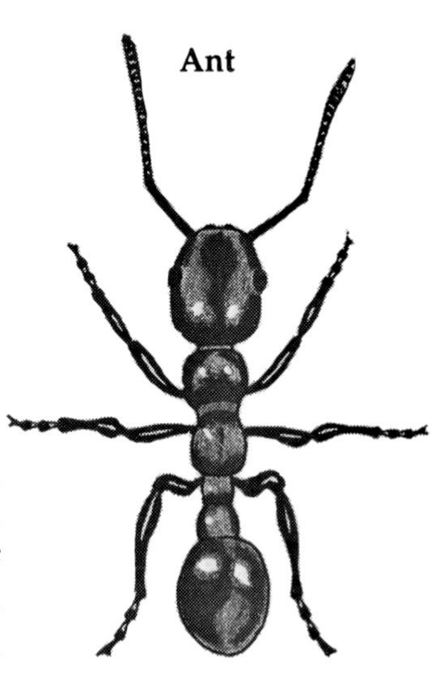

Ants live in colonies (hills, mounds) with queens that lay eggs. The wingless ones we most often see are sterile female workers. Most colonies are located in the soil. All attempts at ant control should be directed at the colony.
Fire ants build small to large hills in pastures, fields and yards. The sting of a fire ant burns much more painfully than that of other ants. Currently, fire ants are generally limited to the Southeastern US. Their further spread will hopefully be limited by climate, but some regions of the Southwest may also be invaded. Mutations to a darker black color have already occurred among the black variety.

All ants have a very narrow junction between their thorax and abdomen. Ant colors vary from black to brown and red.

Life Cycle: 1. Eggs 2. Larvae 3. Pupa 4. Adult, either Queen, worker or male; sometimes there is more than one egg-laying queen per hill. New queens and males are constantly being raised. They are winged and fly away from the hill to form new colonies. An average colony can number 100,000 members and be found as deep into the ground as three feet. For fire ants a good time to apply some controls is during the colder months of the year when ants are less active.

Controls

Biological: *Pyemotes tritici*, or straw mite, is a predatory mite which is a naturally occurring predator of fire ants. They are commercially available. They also attack a wide variety of beetle larvae and caterpillars and many true bugs. Use a special strain of Nc nematode to kill fire ant colonies (see list off suppliers in appendix). Apply 1 - 2 million in a water suspension or on a carrier to the center of a colony at dusk. Wash in with 1 - 2 more quarts of water. Ant lions, the larval stage of brown lace wings or doodle bugs, eat ants but are not yet commercially available.

Cultural: Good aphid control helps control ants in the garden. Interplant spearmint, tansy and pennyroyal in the garden. Bone meal around the plants or juice of hot pepper pods on plants may act as a repellent.

Acute: A mixture of 3 to 4 tablespoons of liquid dish washing soap and ¼ cup of lime per gallon of water can be poured into a colony, 2 to 5 gallons at a time. After first drenching the center and perimeter of the hill, use a stick to probe deeply and uncover the center of the hill. Repeated applications may be necessary for larger hills. Since new queens fly in from other areas, yearly maintenance is necessary. An alternative control is to mix 1 million or less nematodes per 2-5 gallons of soapy drench. In the home, use soap and lime spray or pyrethrin dust or spray directly on the ants. *Ringer Crawling Insect Attack* provides good contact control with no residue. Place sugar or peanut butter as a bait to determine the location of the colony. Boric Acid, which is NaBOH, sodium boron hydroxide, mixed with a mint-apple jelly (trade name *Drax*) is the most preferred product for ants in the kitchen. Notice the similarity to salt, NaCl. In fact, both are found in large surface deposits near one another in California.

Aphids or Plant Lice *Aphidae family, homoptera order (4,000 species)*

Aphid

Aphids are 1/10 to 1/2 inch long pear shaped bugs with or without wings. They are soft-bodied and can be green, yellow or black. They are usually found on the underside of new leaves and along tender plant shoots. Here they pierce plants and suck juices. Ants guard the aphids from other insects and eat the gooey honeydew, which aphids excrete. Winged forms and male aphids usually appear around autumn.

Life Cycle: 1. Egg on leaf 2. Hatch to nymph stage (small adult form) 3. Adults feed on leaves and lay eggs.

Controls

Biological: Ladybug beetles and their larvae, lacewings and their larvae and aphid midges (*aphidoletes aphidimyza*) all eat aphids and aphid eggs.

Cultural: Fall plowing helps kill overwintering aphid eggs outdoors. Interplant mint.

Acute: Hose off plants with water. Once disturbed aphids may not return. Dust wet plants heavily with lime or spray plants with lime and water solution. Use homemade soap spray alone or lime and soap spray together. *Ringer Aphid-Mite Attack Insecticidal Soap* provides good control. *Safer™ Insecticidal Soap* and *Chevron-Ortho Insecticidal Soap* are available. Also newer horticultural oils like *Sunspray* smother aphids and aphid eggs.

Apple Maggots *Rhagoletis pomonella*

These yellow-white legless larvae bore into apples after hatching from eggs laid into punctures made by the parent flies. The mature maggots fall to the ground, pupate and winter over. They can be found throughout the US.

Controls

Biological: Apply Nc nematodes as a soil topdressing in the late summer to early fall to kill maggots as they fall to the ground. Place plastic apples or solid red and yellow balls together in the trees in and near the orchard, smeared with tanglefoot or another sticky substance in the spring and summer to trap adults. Commercial traps are also available.

Cultural: Pick up and use or destroy all apples as they fall.

Armyworms *Euxoa auxiliaris and several other species*

Fall Armyworms *Spodoptera frugiperda and several other species*

This greenish two inch long caterpillar with yellow stripes down its back is actually a type of cutworm feeding near the ground, often on stalks. Its gray moth lays eggs on grasses and corn. The night feeding worms eat that grass until they grow larger, then as their food runs out they "march" to other plants looking for food. The "fall armyworm" has a similar biology but comes in a variety of colors and doesn't leave host plants during the day. Fall armyworm moths fly north from the southern US every spring.

Life Cycle: 1. Eggs laid on leaves 2. Larva feeds on grass and stems on ground at night in the early spring and later feeds on plants and leaves 3. Pupates in soil 4. Adult butterfly

Controls

Armyworm

Biological: Use trichogramma, lacewings, and beneficial Nc nematodes as a soil mulch. Spray or dust Bt or NPV on plants. Handpick in soil near plant.

Cultural: Control weeds, mow grass, interplant strong smelling herbs, fall plow.

Acute: Use Nc nematode and Bt. Spray soap and lime directly on caterpillars. Use commercial pheromone traps for monitoring and low level trapping out of adult stage moth. Homemade light traps using funnel design for night flying moths can also be used.

Bagworms *Thyridopteryx ephemeraeformis*

Bagworm

These pests are "wolves in sheep's clothing", pests that disguise themselves with parts of the arbor vitae, cedar or other shrubs or foliage they eat. Inside the spindle-shaped bag formed of leaf parts that they glue together with silk are dirty brown caterpillars. They drag their bags along as they feed. A horde of bagworms can easily defoliate and kill a tree. Most are found east of the Mississippi River.

Life Cycle: 1. Eggs laid in the bag winter over and hatch in early spring 2. Larva 3. Pupa in late summer 4. Adult butterfly

Controls

Biological: With light infestations handpicking bags is effective. Use Bt.

Acute: For heavier infestations spray Bt on foliage. Avoid chemical contact poisons which kill naturally occurring parasitic wasps. *Malathion*, for example, has a residency period of 30 days and a negative effect on wasps like trichogramma due to fumes and vapor. This invites reinfestation.

Bean Leaf Beetles *Cerotoma trifurcata*

These are ¼ inch long and reddish with 4 black spots and a black head. Their eggs are laid on leaves. The larvae enter the soil and feed on roots. Adult beetles feed on leaves.

Life Cycle: 1. Eggs on leaves 2. Larvae enter soil 3. Pupa 4. Adults feed on leaves

Controls

Biological: Ladybug beetles and lacewings eat eggs. As an experiment for killing larvae and emerging adults, mix beneficial nematodes into seed furrow and use as a mulch around plants. Handpick bugs off plants and destroy.

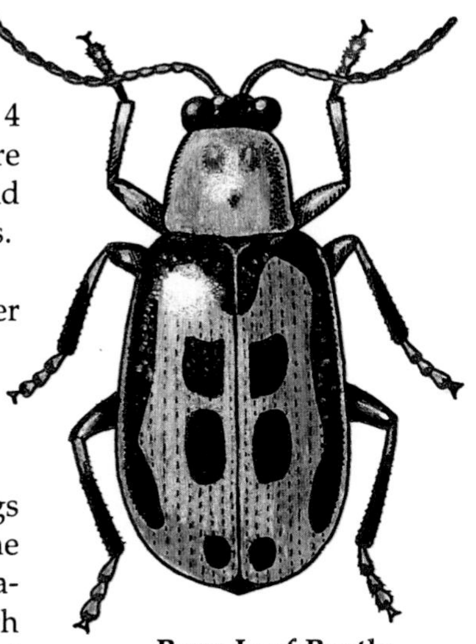

Bean Leaf Beetle

Cultural: Interplant potatoes, garlic, cloves and radishes along with beans. Fall plowing helps destroy eggs.

Acute: Spray garlic, onion or pepper extract to repel or soap and lime spray on beetle to irritate. Spot spray rotenone or a pyrethrin product.

Blackflies *Simulium damnosum and other species*

These 1⁄25 to 1⁄5 inch gnats are a nuisance to gardeners. They bite around the head and arms, suck blood, then lays eggs in moving, well oxygenated water. The eggs hatch into bowling pin shaped black maggots that attach to stones and bottom debris in streams. Usually there is only one generation per year, but even a small stream can produce millions of blackflies each spring. Biting species can have 3-4 generations per year.

Controls

Biological: For control the following spring, apply Bti rings upstream or spray Bti in the late summer or early fall. Bti is a special strain of Bt which only affects blackfly and mosquito larvae.

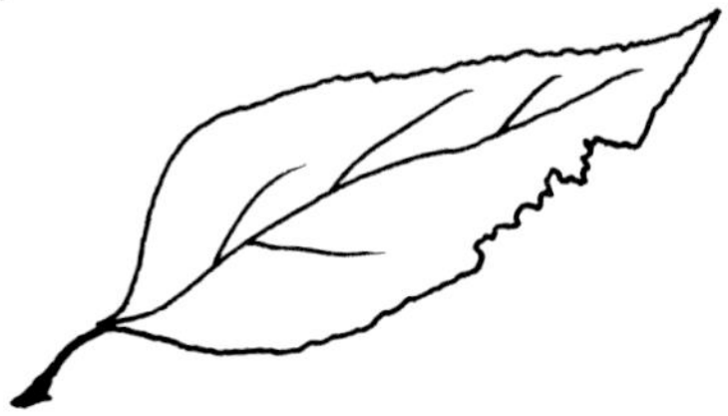

Leaf notches

Black Vine Weevils *Otiorrhynchus sulcatus*

This shy night feeding weevil eats notches on the sides of the leaves of yew tree (taxus), rhododendron bushes, azaleas. and other ornamental shrubs. It lays eggs in soil. The eggs hatch into white legless grubs which feed on roots and kill plants. During the day weevils remain well concealed in mulch and leaf litter on the ground. The leaf notches are the surest signs of this serious pest.

Controls

Biological: Use insect eating nematodes in a mulch around bushes. Hoe in around roots. Applying tanglefoot around trunk of the bushes or trees will prevent adults from climbing and eating leaves.

Cultural: Be sure to buy uninfested nursery stock. Prune lower branches that are in contact with the ground.

Blister Beetles *Epicauta pestifera (several other species)*

These are solid black flying beetles, sometimes black with yellow stripes. They are ½ inch long with long legs and a narrow neck. They cause skin to blister if handled. They eat leaves, fruit and stems, but their larvae are all beneficial and prey on grasshopper eggs, so you might want to leave them alone.

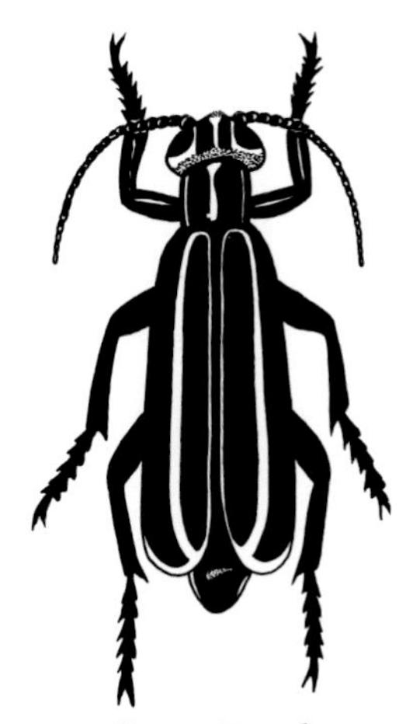
Blister Beetle

Life Cycle: 1. Eggs laid in soil 2. Larvae in soil 3. Pupa in soil 4. Adult beetle

Controls

Biological: Handpick with gloves or use broom to move large populations away from your garden. Experiment with Nc nematodes in hay pastures and grassed yards for larval stage control.

Acute: Dust wet plants heavily with lime or use lime spray to cause them to leave the garden. Use soap and lime spray if all else fails. Spot spray a pyrethrin or rotenone product on plants.

Cabbage Loopers *Trichoplusia ni*

This white striped green caterpillar forms a loop with its body when it crawls. Its nocturnal moth lays single eggs on leaves and is brownish in color.

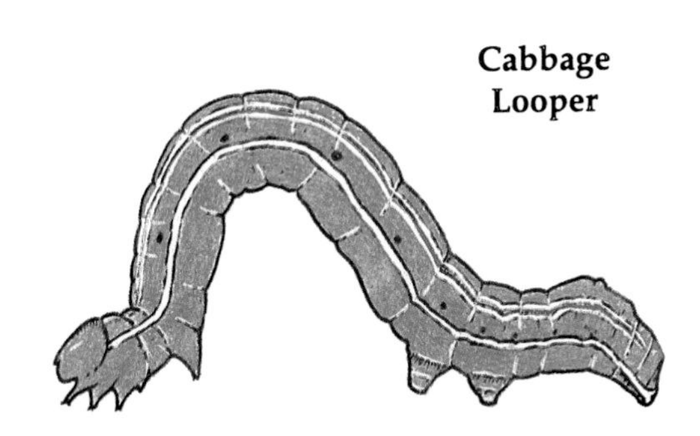
Cabbage Looper

Life Cycle: 1. Eggs on leaves 2. Larva eats leaves 3. Pupa 4. Adult butterfly

Controls

Biological: Release trichogramma and lacewings for preventative control. For quicker control, dust or spray Bt or NPV.

Cultural: Interplant strong smelling herbs to deter butterfly or moth egg laying. Handpick caterpillar off plant. Fall plow.

Acute: Dust wet plant with lime or spray with soap and lime. Spray with Bt or NPV if labeled for loopers. Use Nc nematodes as a spray on plants. Apply late in the day.

Cabbage Root Maggots *Hylemya brassicae*

These legless white maggots with a black "hook" in their heads attack the roots of cabbage, broccoli, cauliflower, radishes, turnips and some other plants. Similar maggots attack carrots and onions. Often their attack will rot the whole root or the plants die. Plants set in the fall can also be attacked. Two generations may occur per year.

Life Cycle: 1. Eggs laid in soil crevices 2. Larva (maggot) in plant roots 3. Pupa in soil 4. Adult fly hatches in early spring

Controls

Biological: Apply beneficial Nc nematodes at planting in seed furrow or around transplants.

Cultural: Crop rotation, never plant susceptible plants within flying distance of areas planted in the last two years.

Acute: Use Nc nematodes or lime spray or dust on ground around plants to deter egg laying.

Chinch Bugs *Blissus leucopterus*

This ⅙ to ⅕ inch black bug with white, red or brown wings feed on grasses and corn. The adults hibernate in grass then lay eggs in the soil in the spring. These hatch into reddish wingless nymphs. In about 30-40 days they turn into adult bugs and produce another generation. The grains and grasses they attack look brown and dried out, caused by the nymphs and adults sucking sap.

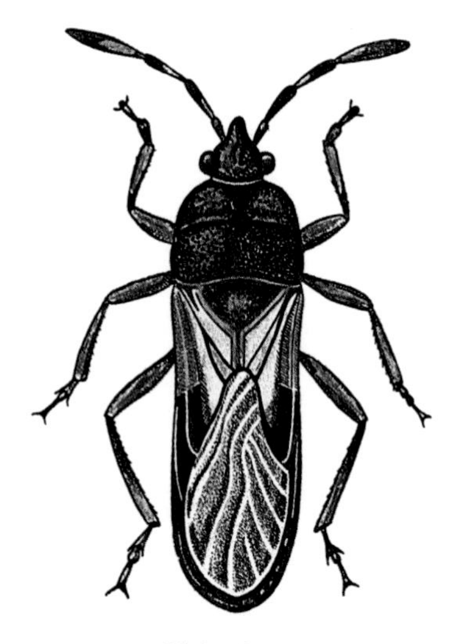

Chinch Bug

Life cycle: 1. Eggs in soil 2. Nymph (small adult) 3. Adults that winter over in grass

Controls

Biological: As an experiment use beneficial Nc nematodes applied as a soil top dressing or mulch. They may attack adults and nymphs as they hide on the ground or in grass. Keep ground moist. Ladybug beetles and lacewings provide some control.

Cultural: Use fall plowing. Clean up weed and brush where adults winter over.

Acute: Use soap spray to kill bugs or spray garlic, onion and pepper extract to repel them. Use soap spray on heavily infested lawns.

Cockroaches *Blattella germanica*

There are four main North American species of this house pest. They eat everything that humans eat plus some of what they can't or won't, such as paper. They thrive in moist, warm and dark environments.

Controls

Cultural: Dry houses in colder climates with floors that are crevice-free can deter cockroaches. In other situations, good sanitation is a must. Clean up food and dishes at night and take care of any damp areas. Keep garbage cans and food sealed in metal or thick plastic. Seal the house with tight fitting screens and doors. Seal around pipes leading outside with caulking material. Seal around window screens and doorways if necessary.

Acute: Use boric acid powder, trade name *Roach Kil*™ or *Roach Proof*™ or a similar product. Most people overapply this product. A very light fine dusting near cracks and crevices where roaches walk is all that's needed. Boric acid bait stations, trade name *It Works,* are available in a plastic child proof enclosure. A boric acid product to use in mop water, called *Mop-Up,* is commercially available.

Keep boric acid powder products stored away from children. Do not use the powder where pets can get it on them. **Boric acid is a poison.** It is very long lasting and should be the number one product for roach control in the house. See Additional Reading Section in the back of this book for more information on roach control. See especially "IPM for Home and Garden" and *Least Toxic Roach Control* from the publishers of "IPM Practitioner" and the "Common Sense Pest Control Quarterly."

Codling Moth Larvae *Laspeyresia pomonella*

This pink, or white tinged with pink caterpillar is the proverbial worm in the apple. Without some kind of control, most apples get them.

Life Cycle: Two or three generations per year, occurring every 5-8 weeks. 1. Emerging moths lay eggs on leaves, twigs and buds. 2. Larva enters the apple, grows, emerges and climbs down tree trunk 3. Pupates either under loose bark, in trash

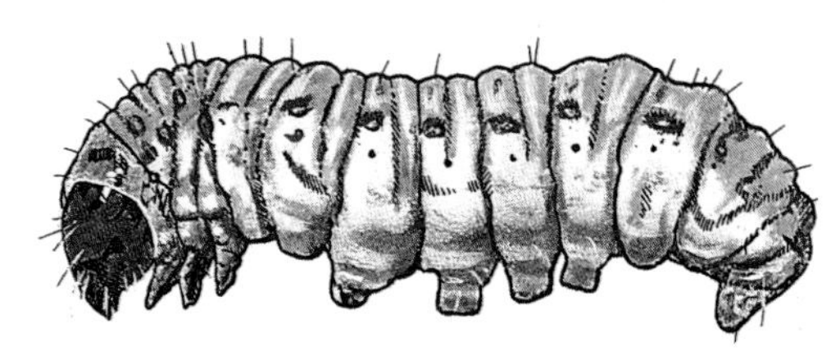

Codling Moth Larva

on the ground or other protected area. The third generation larva winters over, then pupates in early spring. 4. Emerging moth, butterfly, lays eggs. Moth emergence is extended throughout the spring, summer and fall, but has three distinct peaks.

Controls

Biological: Timed release of trichogramma at the three peaks of egg laying. Begin spring release at petal fall, then release 3-8 weeks later in summer, and 5-8 weeks after that in early fall. Timing releases is important, so it helps to monitor traps. Spray wet tree trunks and underneath trees with Nc nematodes. A late winter spray is most effective before the larvae pupate. Respray three weeks after petal fall, then again after another four weeks or whenever you collect fresh larvae in the cardboard trap. Wrap trunks with several thicknesses of corrugated cardboard one foot or so in length. Remove larvae weekly. Gently scrape loose bark off tree each spring and fall and destroy.

Cultural: Fall cleanup, remove fallen apples, trash and leaves and burn or compost leaves. Interplant clover as a ground cover, and flowers, such as daisies or other field flowers. Plant only disease resistant apple varieties like "Liberty" and "Freedom" that are for your geographic area.

Acute: No acute spray, chemical or otherwise is effective once the worm is in the apple. Spray codling moth virus at petal fall, then at intervals determined by trap monitoring of adult moths. The object is to kill young larvae before they enter apples. Use lime and soap spray weekly on tree trunk and around base on larva. Spray entire tree with soap and lime spray before leaves appear and buds open in late winter (dormant stage spraying).

Colorado Potato Beetle

Colorado Potato Beetles

Leptinotarsa decemlineata

This round ⅜ inch yellow beetle with black stripes eats potato leaves and lays yellow-orange egg masses on the undersides of the leaves. The ½ inch reddish, humpbacked, legged grubs chew leaves and have black dots on their sides.

Life Cycle: Several generations can be completed per year. 1. Eggs on leaves 2. Larva feed on leaves 3. Pupa in soil 4. Adults emerge after 10 days and eventually hibernate in the soil during the winter.

Controls

Biological: Release beneficial insect *Edovum puttleri*, lacewings and ladybug beetles which eat eggs. Handpick eggs, larva and adults off plants. Use *Ringer Colorado Potato Beetle Attack* (Bt-sd) and others for control of early larval stage only.

Cultural: Interplant green beans with potatoes. Fall plow to kill adults.

Acute: Spray garlic, onion and pepper juice as a repellent or soap spray directly on the beetles to irritate them. Use soap and lime spray on larvae to kill them by dehydration.

Corn Earworms *Heliothis zea*
Cotton Bollworms *H. zea*
Tobacco Budworms *H. virescens*
Tomato Fruitworms *H. zea*

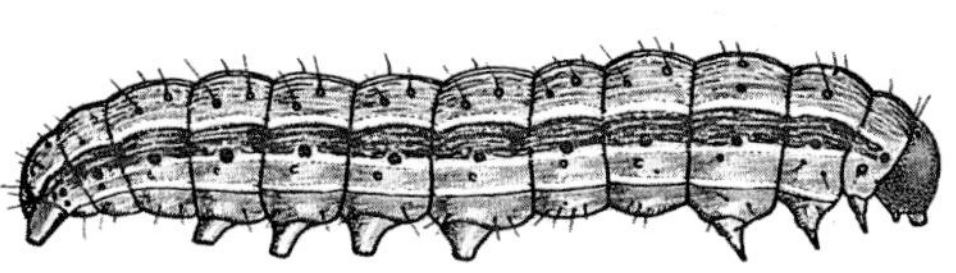

Corn Earworm

The color of these 2 inch caterpillars varies from white, brown, and blackish to red with stripes and hair tufts. They actually attack a variety of plants and have been called the tobacco budworm, cotton bollworm and tomato fruitworm.

Life Cycle: Several generations occur yearly. 1. Eggs are laid by the thousands on a variety of plants including corn, beans and tomatoes 2. Larvae hatch and feed on leaves, then corn silk and kernels of the ear. 3. The caterpillars pupate in the soil 3-5 inches deep, emerging after 10-25 days. 4. Adult moth, the final generation wintering over in the pupal stage to emerge the following spring.

Controls

Biological: Release trichogramma and lacewings. Spread Nc nematodes on the ground as a mulch. Spray Nc on silks or inject onto silks with a medicine dropper at regular intervals. Spray Bt or NPV on leaves and silks. Handpick worms off leaves.

Cultural: Fall plowing destroys pupae in the soil. No lights should be left on near the garden at night to attract the moths.

Acute: See above for Nc, Bt and NPV, spraying weekly for complete control or as needed.

Cowpea Curculios (Bean Weevils) *Acanthoscelides obtectus*

These are small black beetles with a snout. They lay eggs in pods. They are usually found on southern peas and lima beans.

Life Cycle: 1. Eggs laid inside pod 2. Larva feed in pod 3. Pupate in pod 4. Adult emerges

Controls

Biological: As an experiment try nematodes as a soil mulch around plants as a possible control. A praying mantis may eat several adults. Use nematodes around over wintering sites of adult stage, such as wood piles or cracks and crevices in the greenhouse.

Cultural: Use fall plowing and interplant garlic, cloves, potatoes and radishes. Place a sheet or cloth beneath plants and shake gently. Weevils should fall on the sheet. Gather and destroy by crushing or dropping in alcohol. Dry seeds thoroughly and turn or shake once a week, leaving an opening for adults to escape. They will look for a less disrupted resting place.

Acute: Use garlic, onion and pepper spray on pods to deter egg laying. Use lime and soap spray on adult weevils as an irritant.

Cucumber Beetles
Striped *Acalymma vittatum (Syn. Diabrotica vittatum)*
Spotted *Diabrotica undecimpunctata howardi*

Spotted Cucumber Beetle

These are green or yellow beetles with three black stripes or 12 black dots They winter over, then emerge to spread diseases and eat leaves. They feed on cucumbers and related plants, such as melons and squash. The larva are ½ inch long with a dark head.

Life Cycle: Two generations may occur per year in warm climates. 1. Eggs around the bases of the plants and in cracks in the soil. 2. The legged larvae feed in the stems and roots. The larval stage is also called corn root worm. 3. Pupate in soil 4. Adult beetle

Controls

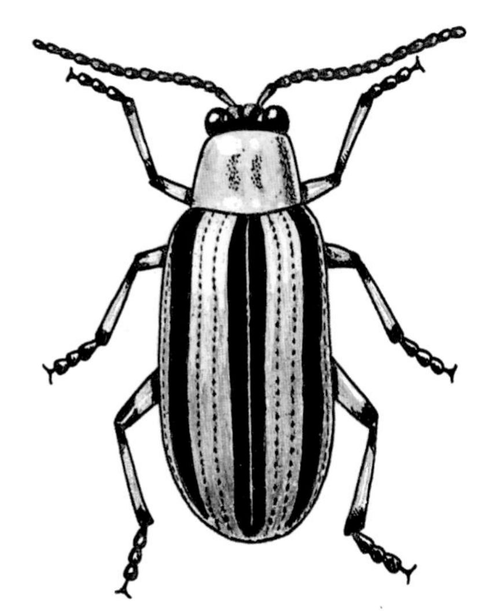
Striped Cucumber Beetle

Biological: Lacewings and ladybug beetles eat eggs. Apply beneficial Nc nematodes as a mulch around plants to kill adults. Apply around roots of transplants or in seed furrow at planting time to kill larvae. Handpick bugs.

Cultural: Interplant radishes, fall plow and use crop rotation.

Acute: Spray plants with lime to deter beetle feeding or soap and lime spray directly on the beetle to irritate and possibly kill them. Use pyrethrin or sabadilla dust.
Caution: sabadilla dust may cause allergic reaction if inhaled. Use dust mask. Certain rotenone products are useful.

Cutworms *Noctuidae family, lepidoptera order (various species)*

There are several species of cutworms. Some are glassy green, others are whitish, brown or black. Some are striped or banded. They feed on a variety of garden plants, including corn, beans, cabbage and numerous others. Many cutworms cut off plants at the base of the stems feeding little, but doing much damage, while many others climb up plants and strip leaves. All these caterpillars tend to feed and move on the soil surface at night. In the day they often lay curled up on or near the soil surface like a "c" near the plants they attack. Some cutworms start behaving like armyworms when food is exhausted.

Life Cycle: In the south several generations occur per year 1. Eggs are laid on plant stems or on bare ground near plants. 2. Larva or worm stage, wintering over buried in the soil 3. Pupate in the spring 4. Adult moth. Some species overwinter in pupal stage.

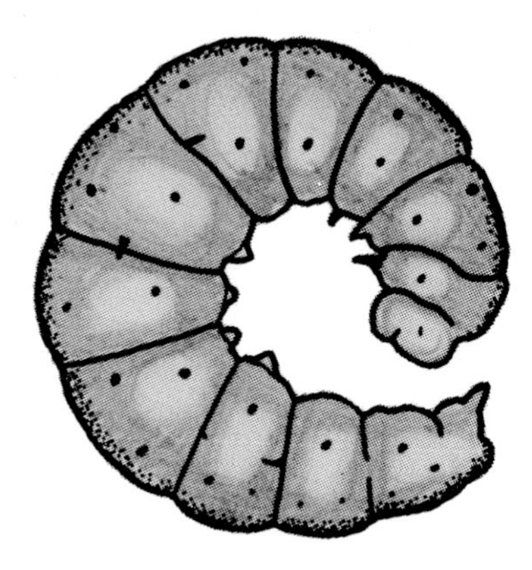
Cutworm

Controls

Biological: Trichogramma and lacewings attack eggs. Apply Nc beneficial nematodes as a mulch around plants. Handpick from soil around plants.

Cultural: Use fall plowing or plow repeatedly before planting. Wrap individual foil or paper collars around stems of plants. Turn off lights near garden at night. Place a piece of sandpaper or a piece of roofing shingle on the ground around plants. The caterpillars will not crawl across the rough surfaces.

Acute: Spread Nc nematodes. Bt spray is effective for strippers. Use Bt spray on the ground and stem to give some help in controlling stem cutters.

Diamond-back moth larvae *Plutella xylostella*

This is a thin green ⅕ inch long caterpillar with a black head. It hangs from a silken thread and eats holes in plant leaves. The moth is gray to yellow in color with a diamond shaped area formed at the tail end of the wings when the wings are folded and it is viewed from the side.

Controls
Same as for imported cabbage worm

Earwigs *Forficula auricularia*

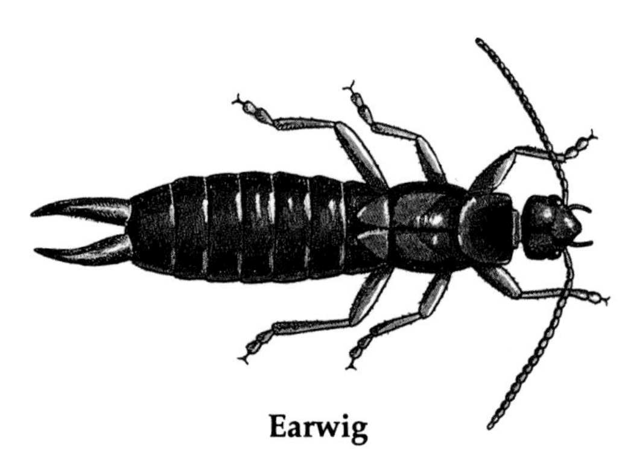

Earwig

These ¾ inch brownish black wingless insects look scary, but do not pinch people with the forceps-like pincers on their abdomens. Since many prey upon aphids and other pests, they are more beneficial than harmful. They like confined areas, are active at night and feed on decayed plant material. Because they kill a pest, clean up the decay it has caused and hide there, we often accuse them of doing the damage to the plant. It would be a mistake to over-react and spray for them. Adult earwigs tend the eggs laid in the soil, then watch over the young nymphs like ducks watching their young until they are able to feed themselves.

Life Cycle: 1. Eggs winter over in soil 2. Nymph stage 3. Adult

Controls

Cultural: Trap them in rolled up newspaper or something else they can hide in. Move to another area. They seek cool protected areas during the day.

Acute: Use insecticidal soap spray directly on the adults.

Eastern Tent Caterpillars *Malacosoma americanum*

These dark caterpillars have a white stripe on their backs and yellow and brown lines and blue spots on their sides. The tent-like webs they spin may be a more distinctive characteristic. They eat the leaves of a variety of fruit and shade trees, except for evergreens. Tent caterpillars widely occur east of the Rockies in the US and southern Canada.

Life Cycle: 1. Eggs, laid in masses that encircle twigs, hatch in the spring 2. Larva eats leaves 3. Pupae winter over in trash or bark 4. Adult moth

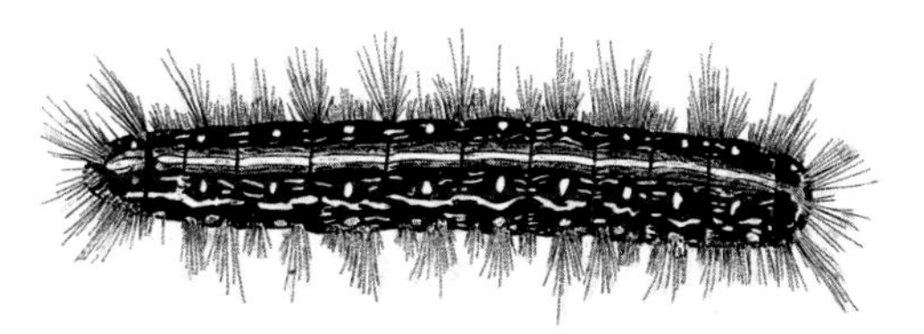

Eastern Tent Caterpillar

Controls

Biological: Release trichogramma and lacewings. Spray Bt on foliage when caterpillars are feeding on leaves.

Cultural: Good fall clean-up of leaves, twigs limbs and fallen fruit and nuts. Remove webb tents and kill all caterpillars. Use strong stream of water to tear up tents high up in tall trees. Interplant flowers and clover.

Acute: Use Bt spray on leaves or soap and lime spray directly on caterpillars.

European Corn Borer *Ostrinia nubilalis*

These one inch gray-pink caterpillars with brown spots are found boring in corn or a variety of grasses.

Life Cycle: Several generations may occur per year. 1. Egg masses on undersides of leaves 2. Small larvae that feed first on leaves, then bore into stalks 3. Pupa 4. Nocturnally active adult moths, pupa winter over in bore holes

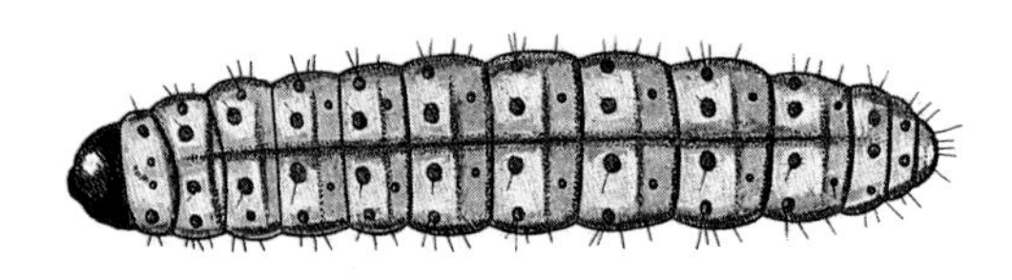

European Corn Borer

Controls

Biological: Trichogramma, ladybug beetles and lacewings all attack eggs. Squirt or spray Nc nematodes into bore hole. Bt can be sprayed before borers enter the stalk, but this has limited effectiveness.

Cultural: Fall plowing destroys overwintering forms. Resistant corn varieties are available.

Acute: Cut a slit in the stalk and remove borer. Use light traps and pheromone traps for monitoring and low level trapping out.

Fleas *Ctenocephalides felis. canis.*
Ticks

The wingless adult fleas are blood-sucking parasites of animals. Their eggs are deposited in dirt, dust, the bedding of their host or carpet.

Life Cycle: Can vary from 2 to 12 weeks, and can be as long as 2 years 1. Eggs 2. Larva 3. Pupa 4. Adult flea

Controls

Biological: Beneficial Nc nematodes attack flea larvae in turf. Keep it moist. *Hunterellus hookeri* is a tick parasite. Use a white canvas drag sheet to monitor tick levels.

Cultural: Inside vacuuming sucks up eggs, larvae and pupae at the same time it removes flea food. Clean pet bedding frequently. Clean up animal refuse outside.

Acute: Use pyrethrin-based flea collars, pet shampoos and soaps on dogs. Sprinkle pyrethrin-based flea powder on carpet followed by vacuuming and/or rug shampooing. Remove and dispose of vacuum bag immediately after vacumming. In the yard, spray or dust a pyrethrin or rotenone product. Use on animal refuse. **Caution:** pyrethrin products should not be used on cats due to allergic skin reaction. Use rotenone or pyrethrin spray outside on heavily tick infested areas such as picnic sites. A parasite of ticks should be released for more permanent control. See *Least Toxic Flea and Tick Control* from "IPM Practitioner".

Flea Beetles *Phyllotreta striolata*

A jumping beetle, ¼ inch or less, sometimes with black and yellow stripes. They eat tiny holes in young leaves. Larvae feed on seeds and roots.

Life Cycle: 1. Eggs in soil 2. Larva in soil 3. Pupa 4. Adult beetle

Controls

Biological: Apply beneficial Nc nematodes in seed furrow and as a mulch around plants.

Cultural: Use good weed control and repeated early spring cultivation along with fall plowing. Interplant mint.

Acute: Use garlic, onion and pepper spray as a repellent. Heavily dust wet plants and the wet ground around plants with lime, lime spray or soap and lime spray.
Caution: do not overuse lime. It could change the soil Ph.

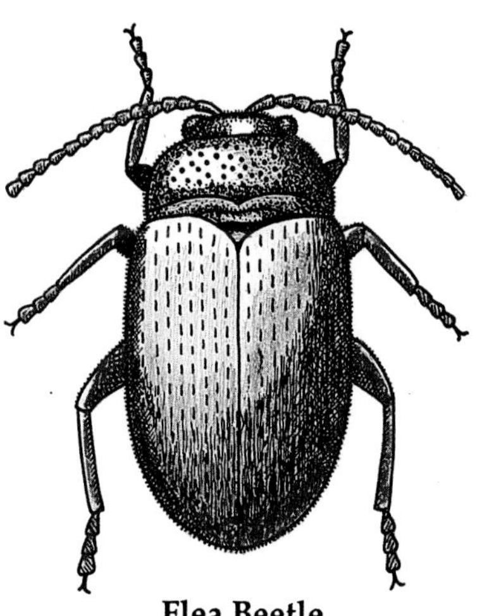

Flea Beetle

Flies, (including house fly and other filth fly species) *Musca domestica*

Various species of these flies all require fresh excrement from a warm-blooded animal to complete their life cycle. Needless to say, they can spread disease. Others bite and suck blood or suck tears.

Life Cycle: In hot weather, cycle can take only a couple of weeks. 1. Cylindrical eggs in fresh manure 2. Legless maggots with a black hook on their heads 3. Pupate in a drier area, pupa has a brown, leathery shell 4. Adult fly emerges in several days

Controls

Biological: Introduce fly parasites each year on a weekly basis near fresh and dry manure. These parasites deposit their eggs in fly pupae. They are commercially available for use around dairies, chicken houses, feed mills, stables, barns, etc.

Cultural: Clean out fresh manure every three days and allow to dry. Dry manure harbors fly parasites so leave about 8 inches of the dry manure base. Eliminate water leaks or continuously moist areas.

Acute: If adult fly populations get out of hand, use pyrethrin aerosol spray or fly traps. Do not use around fly parasite breeding areas. Keep pyrethrin away from water and animal feed and bedding. Do not spray inside the barn.

Grasshoppers *Acrididae family, orthoptera order (various species)*

Various species of 1 to 2 inch grasshoppers chew on almost all wild and cultivated plants. Many are gray-brown-green.

Life Cycle: 1. Egg masses laid in soil 2. Wingless nymphs hatch the next year 3. Winged adult develop by summer

Controls

Biological: Praying mantis may eat grasshoppers. The beneficial protozoan *Nosema locustae* controls most species of grasshoppers and some crickets. It is sold as *Grasshopper Attack* and several other brand names. If you live in an area plagued by grasshoppers, you will get quickest benefits by applying it as a preventative by early summer every other year when the nymphs are small. The effect of this treatment will be optimal the next summer.

Cultural: Fall plowing destroys eggs, but grasshoppers often lay eggs in rough, unplowable areas.

Acute: Spray soap directly on the grasshoppers. As an experiment, mix some nematodes with the soap spray and apply late in the evening or at night.

Leaf Rollers *Lepidoptera order (various species)*

Several species of caterpillars, all spin a web in leaves and roll leaves to pupate in.

Life Cycle: 1. Eggs laid on stems 2. Larva or caterpillar stage 3. Pupa or cocoon stage 4. Adult butterfly

Controls:

Biological: Release lacewings or trichogramma to destroy eggs. Spray Bt on foliage to kill caterpillars.

Cultural: Cut off and destroy infested leaves. Interplant strong herbs to discourage egg laying.

Acute: Bt spray on plant leaves. Use NPV or soap and lime spray directly on caterpillars.

Leafhoppers *Graphacephala coccinea (several other species)*

These slender ⅛ inch long wedge-shaped bugs suck plant juices, kill leaves and spread virus diseases of plants. There are several species. Some are all green. Others are green with red and white markings. All of them hop or fly when disturbed. Often they sit and feed on the undersides of leaves. They are pests on a variety of plants including apples, beets and potatoes. Dragonflies eat leafhoppers.

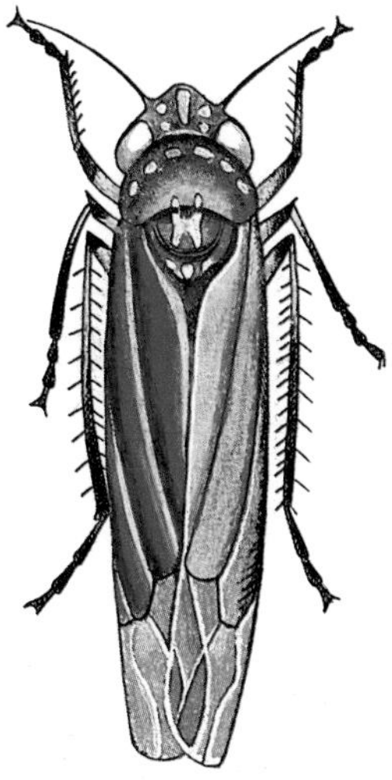

Life Cycle: 1. Eggs on plants, winter over 2. Nymph (small adult) 3. Adult, some hibernate in trash, others migrate north in spring after wintering over in the south.

Controls

Biological: Lacewings eat eggs.

Leafhopper

Cultural: Use fall plowing.

Acute: Use insecticidal soap spray early in the morning when the insects are less active.

Lesser Corn Stalk Borers *Elasmopalpus lignosellus*

The green, blue or brown banded caterpillars tunnel into corn stems, usually near the soil surface and wriggle wildly when held. The caterpillar exits the stalk through a tube-like silken web and pupates in the soil. This pest may occur in other regions and damage other plants, but is worst in Southern corn.

Life Cycle: 1. Eggs on leaves 2. Larva 3. Pupa in soil 4. Butterfly

Controls

Biological: Beneficial Nc nematode as a mulch or inject into the borehole. Release trichogramma, ladybugs and lacewings.

Cultural: Fall plow. Make a small cut on the stalk near bore hole and remove borer. Use crop rotation.

Acute: Use Bt and NPV sprays to provide some control before caterpillar enters stalk.

Mealybug *Pseudococcus adonidum*

Adults are elliptically shaped with short spines, actually a part of their waxy white covering. Females have no wings, do not move and no legs are visible. The males do not feed in this stage, but transform into a two-winged flier that mates with the immobile female, then dies. Eggs are laid within the waxy case and hatch into six-legged bugs with smooth bodies. These disperse, "stick" to the plant and begin sucking juices. Soon afterward, they secrete the white covering and after that only move sluggishly. They attack a wide variety of soft-stemmed foliage plants under greenhouse conditions. Here the life cycle takes about one month.

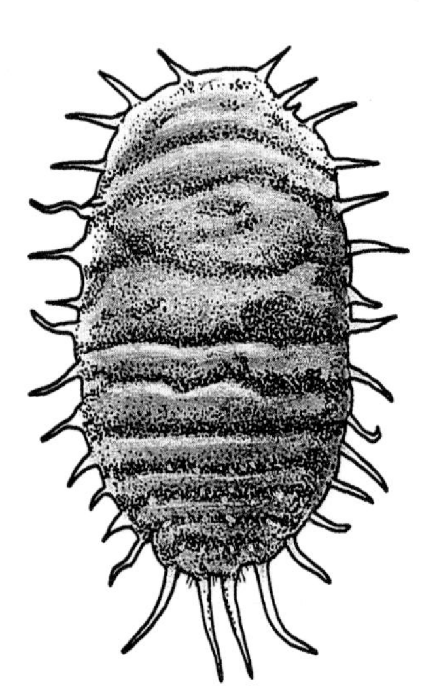

Mealybug

Life Cycle: 1. Eggs laid in protected areas under bark 2. Nymph (small adult) 3. Adult, ¼ inch in length.

Controls

Biological: Both the *Cryptolaemus* ladybug (not the common one) and *Pauridia* parasites are commercially available for use.

Cultural: Quarantine new potted plants for 30 days and examine weekly for mealybugs. In early spring spray dormant stage with soap.

Acute: Wash off with high pressure water, but be careful not to damage plants. Spray soap or touch individual insects with alcohol on a cotton swab. It may take repeated application.

Mexican Bean Beetles *Epilachna varivestis*

Light brown beetles with 16 black dots on their backs resemble ladybug beetles. They skeletonize bean leaves and lay yellow egg clusters on leaf undersides. Their yellow-orange larvae also feed on leaves.

Life Cycle: 1. Eggs on leaves 2. Larva eat leaves 3. Pupa 4. Adult beetle

Controls

Biological: Release parasite *Pediobius foveolatus*. Handpick bugs and eggs.

Cultural: Interplant potatoes, garlic, cloves, turnips, marigolds and radishes. Fall plowing helps control populations.

Acute: Garlic, onion and pepper juice spray will deter feeding. Use lime and soap spray as an irritant or pyrethrin to kill. Rotenone is labeled for Mexican Bean Beetles.

Mexican Bean Beetle

Moles

These furry, warm-blooded animals live underground. They tunnel, searching for grubs and worms. They don't eat roots, but the tunneling damages gardens and the yards.

Controls

Biological: You may want to plant four or five castor bean plants or "mole plant", *Euphobia lathyris*, in or near your garden. Both plants are somewhat poisonous. Do not allow small children to eat their leaves or stems.

Cultural: The major cause of mole activity in lawns and gardens is an excess of pest insects like white grubs. Control the grubs and the moles will leave the area.

Acute: Use commercially available spring loaded mole traps.

Mosquitoes *Culex pipiens (several other species)*

Female mosquitoes digest a blood meal, then lay eggs in water or on a surface that is regularly flooded. The water of mosquito development must be still. Eggs hatch into aquatic larvae (wrigglers) that feed on bacteria, then turn into pupae in the same water. In hot weather the adults emerge in several days. For certain species there is only one generation per year, but there are multiple generations of some species in most of North America.

For information on area-wide program see literature from "IPM Practitioner".

Controls

Biological: Female mosquitoes usually choose shallow bodies of water lacking fish or other insects which can prey on their young. If you have ponds or lakes with mosquito larvae, you may want to stock mosquito-eating fish, *Gambusia affinis*. Keep the shore line steep so fish can get to the larvae. You can also treat with rings of Bti, a special strain of Bt that affects the larva only. The rings float and slowly release Bti bacteria. In North America many larvae grow in small puddles and in "tree holes" in the crotches of trees. These can be sprayed with Bti or soap spray.

Cultural: Encourage purple martins, which eat adult mosquitoes. Build apartment-like bird houses, set up gourds or gallon milk containers near your home. Make entrance holes 2⅛ to 2½ inches, allowing a depth from the hole to the bottom of the inside of 6 inches. Also, drill a small bottom hole for drainage. Houses should be 8 feet or more off the ground. Also, consider using bat houses. Contact Bat Conservation International, PO Box 162603, Austin, TX 78716-2603 for design of bat house. Include SASE for a reply.

Pecan weevils *Curculio caryae*
Chestnut weevils, other nut weevils (similar insects)

These species of weevils have extremely long snouts, as long as their bodies, ⅜ to ¾ inch, and long elbowed antennae parallel to the snout. The females use their snouts to bite holes in developing nuts. Here they lay eggs. The eggs hatch about the same time the nuts fall and develop into legless grubs, "worms". Ten per nut is not uncommon. They eat through the nut meat, then exit and burrow deep into the ground where they overwinter.

Life Cycle: 1. Egg 2. Grub 3. Pupa 4. Adult

Controls

Biological: Spread Nc nematodes as a topdressing under the tree foliage in late summer or when nuts fall.

Cultural: Pick up all nuts quickly after they fall. Freezing or roasting should kill small grubs. Do a good fall clean up of trash and stems.

Pickle worms *Diaphania nitidalis*
Melon worms (similar insect)

The night flying moths lay cluster of eggs on vegetation and on the undersides of cucumbers, cantaloupe, squash and pumpkins. The young black spotted caterpillars may feed on vegetation, but many find their way to fruits and bore into them, growing into a greenish or coppery ¾ inch caterpillar. After about two weeks they desert their burrows and pupate, often rolled inside a leaf. With spring temperatures the moth emerges in a month. The pickle worm occurs east of the Rockies and as far north as New York, but is worst in the South. It spreads yearly from the South, since it cannot winter over in the North. The melon worm is similar to the pickle worm, but has stripes, no spots and feeds more extensively on foliage.

Life Cycle: 1. Eggs laid on plant 2. Larva or caterpillar 3. Pupa within rolled leaf 4. Adult moth

Controls

Biological: Release trichogramma and lacewings.

Cultural: Interplant strong smelling herbs to deter egg laying. Fall plow to destroy pupa in fall. Plant early in the North, before it has a chance to migrate up from the South. To protect your canteloupe, plant squash nearby to attract pickle worms, which prefer squash, then destroy the squash and its worms.

Acute: Weekly spraying of Bt or as needed.

Pickle Worm

Root-knot nematode *Meloidogyne species*

This is a plant-infecting nematode. It causes cell decay and root galls to form by releasing toxins and bacteria as it feeds in the roots. Galls are large round formations on the roots, not to be confused with legumen nodules. Nematode galls split open easily and decayed plant tissue is obvious. Plants infected this way wilt easily and appear stunted.

Life Cycle: 1. Eggs inside gall 2. Juvenile inside gall 3. Adult emerges from gall into soil

Controls

Cultural: Thickly interplant the French variety of marigolds (*Tagetes patula*) or totally plant the infested area with marigolds one season and plow them into the ground in the fall, stems, roots and all. Marigold roots release a substance that is toxic to these nematodes. Use compost near susceptible vegetables, such as okra and tomatoes. Compost encourages the growth of a beneficial fungus that kills these nematodes in the soil. Subsequently, use good crop rotation practices and interplant marigolds.

Scale *Coccidea family, homoptera order (various species)*

All scale have a biology similar to mealybugs. However, they usually have no white waxy covering, but rather a hard shell of different colors and shapes. Many do not lay eggs, but bear live nymphs.

Life Cycle: Out-of-doors cycle may be up to a year; indoors it is about a month 1. Eggs or live born 2. Nymph (several stages) 3. Adults

Controls

Biological: For softscale or black scale, *Metaphycus helvolus*, a wasp parasite, is available. For hard, red scale, there is *Aphytis melinus*. The *Chilococorus nigritis* ladybug is a scale predator. Regular ladybug beetles also will prey upon some scales.

Cultural: Gently scrape scale off plants or touch with cotton swab soaked in alcohol. Repeated application every 3 or 4 days may be needed or until scale dies.

Acute: Early spring before buds open use soap and lime spray or soap spray any time if plant will not adversely react to soap or alcohol.

Slugs and Snails *Mollusca phylum*

Snails scrape small holes in foliage as they feed. They lay masses of eggs.

Controls

Biological: The predatory snail, *Ruminia decollata,* is commercially available. Check restrictions on interstate shipment first. Handpick.

Cultural: Band citrus trees with a copper sheath that is commercially available.

Acute: Place shallow pie pans containing stale beer in the soil so their rims are at ground level to trap. Place boards in the garden to trap. Snails seek cool, protected places during the day and slugs crawl under boards during the day. Commercial traps using the same principal are available.

Spider Mites *Tetranychus urticae*

These non-insect pests are $\frac{1}{50}$ - $\frac{1}{150}$ of an inch long. Spider mites are orange, brown or green with eight legs and spin delicate webs. Their adults and nymphs suck juices and turn leaves yellow, silver or speckled.

Life Cycle: 1. Egg on plants and leaves 2. Nymph (small adult) 3. Wingless adult

Controls

Biological: Predatory mites are the most effective, along with ladybug beetles and lacewings.

Cultural: Quarantine new potted plants.

Acute: Spray insecticidal soap, lime spray or a combination of both. Dust wet plants heavily with lime to repel.

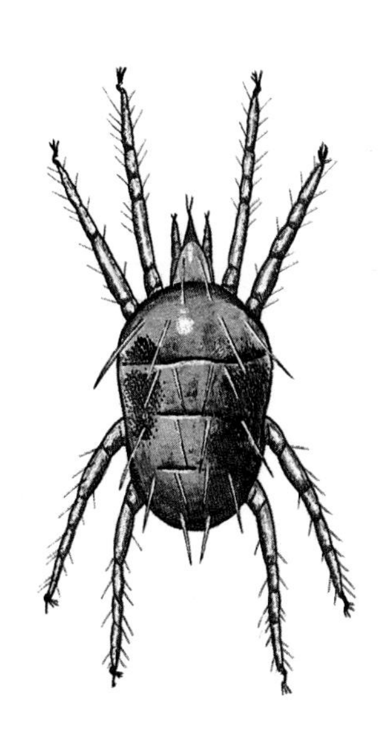

Spider Mite

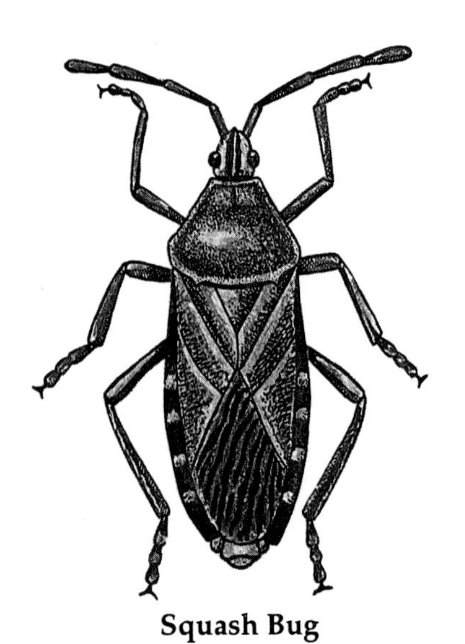

Squash Bug

Squash Bugs *Anasa tristis*

This brown, shield shaped, ⅝ inch long bug sucks plant juices and can be found on leaves.

Life Cycle: 1. Brown eggs on undersides of leaves 2. Nymphs (small adult) 3. Adults winter over in trash

Controls

Biological: Praying mantises eat eggs and nymphs. Handpick bugs and eggs.

Cultural: Interplant radishes and marigolds.

Acute: Use insecticidal soap spray directly on bugs.

Squash Vine Borers *Melittia cucurbitae*

The clear winged wasp-like moths lay single eggs, often near the base of stalks. The caterpillars hatch, then bore into the stalk or leaf stems. They tunnel along, feeding on inner tissues, growing into fat, one inch long borers with wrinkled skin. The borers push brown refuse out of their entrance holes. These borers occur in the Eastern US, attack squash, pumpkins, cucumbers and others.

Life Cycle: In the South, two generations may occur before wintering over as pupae 1. Eggs laid in the soil 2. Larva or caterpillar stage, white with dark head 3. Pupate in soil after about a month as a caterpillar 4. Adult moth

Controls

Biological: Release trichogramma to attack borer eggs. Use Nc nematode as a mulch around vines.

Cultural: Interplant garlic or onion. Use fall plowing to destroy pupa. Harvest fruit as soon as full grown. Use pheromone traps to monitor for adult moths.

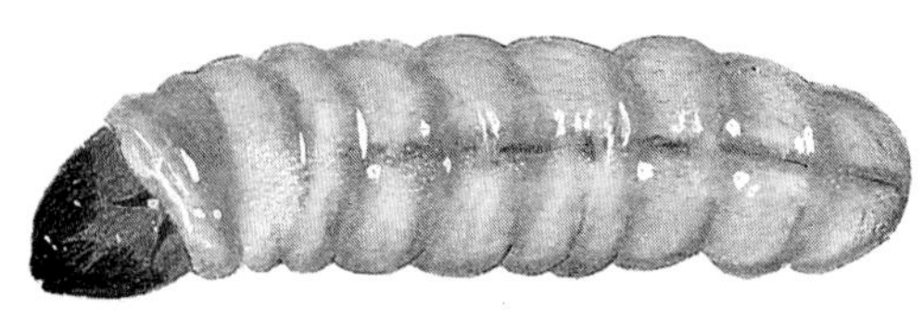

Squash Vine Borer

Acute: Try injecting Nc nematodes into the borehole with a medicine dropper. Slit stem to remove borers and pack dirt over the slit. Use weekly spraying of Bt when first vine runners appear.

Stink Bugs *Nezara viridula*

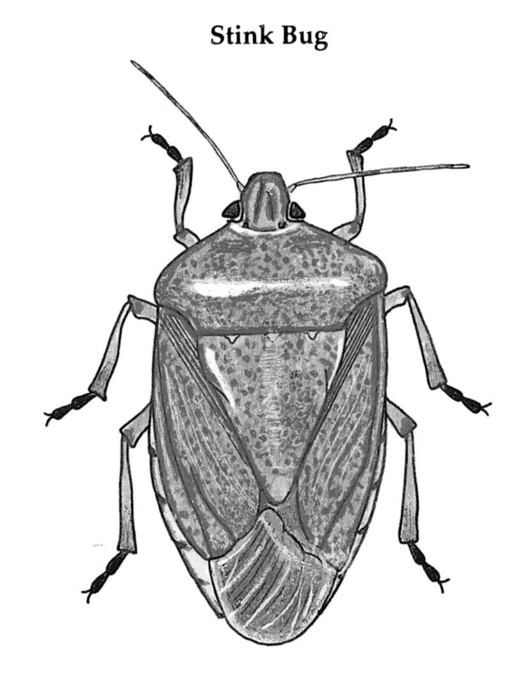
Stink Bug

Some species of these flat ¼ - 1 inch long shield-shaped bugs cause white spots on tomatoes as they suck juices of both plants and other insects. A number of species parasitize pest caterpillars and beetles, so you may not want to kill them if they are not causing a problem.

Life Cycle: 1. Clusters of brown or green eggs on underside of leaves 2. Nymphs (small adult) 3. Adult bug

Controls

Biological: Handpick bugs and eggs off plant. Use parasite *Trissolcus basilis.*

Cultural: Use good weed control, fall plowing and interplant radishes.

Acute: Spray soap and lime spray. Spray pyrethrin.

Strawberry root weevils *Brachyrhinus ovatus*
Strawberry rootworms

These brownish ⅛ inch long weevils have an appearance and life cycle very similar to black vine weevils. The smaller rootworms are the spotted, legless grubs of the weevils. The weevils feed on the upper plant, then lay eggs in the soil. The grubs feed on roots; heavy infestations kill the plant.

Controls

Biological: Apply Nc nematodes as a mulch. A sandy soil makes nematode penetration easier.

Acute: Try handpicking, although the nocturnal adults are very shy. Hoe in mulch with Nc nematodes.

Sweet Potato Weevil

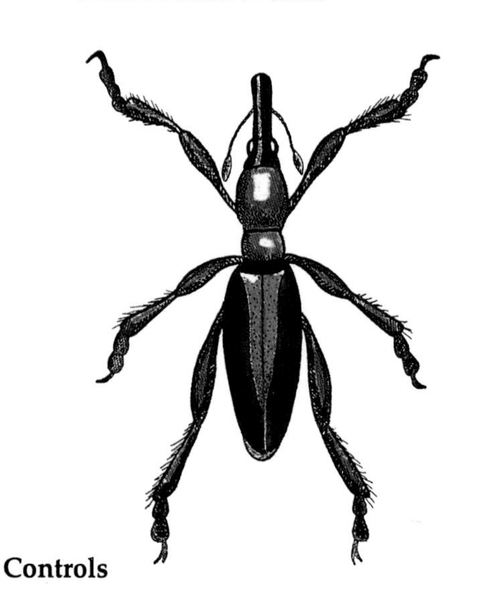

Sweet potato weevils

Cylas formicarius

This ¼ inch ant-like beetle with a snout has a black body with red legs. They feed on foliage and lay eggs in the stem and tubers. The legless grubs hatch and eat down the stem into the potato or roots. The weevil is limited to the South, where it is active year-round and also attacks stored sweet potatoes.

Life Cycle: 1. Eggs in soil or plant 2. Larva 3. Pupa in tuber and stems 4. Adult beetle

Controls

Biological: As an experiment, use Nc nematodes around well mulched plants.

Cultural: Plant only stock from weevil-free areas. If infestation occurs, burn all plants and tubers, then replant after one year in a different location. It is also helpful to notify your county agent of infestation if you live in a state where commercial sweet potato growing occurs.

Acute: Dust adults with rotenone.

Tarnished Plant Bug

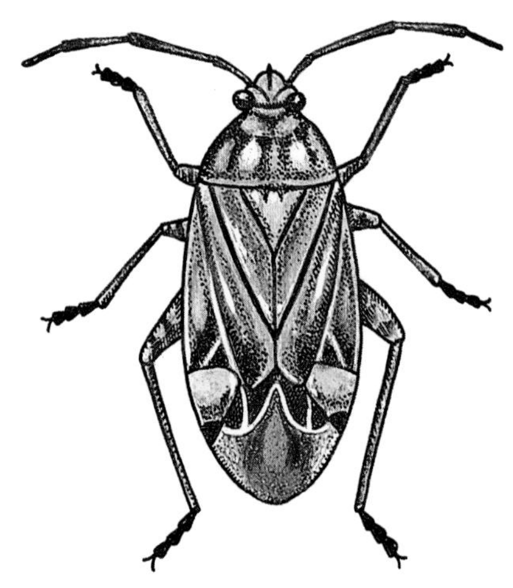

Tarnished plant bugs

Lygus lineolaris

These ¼ inch black and brown bugs and their nymphs suck plant juices. This causes darkened areas. Eggs are laid on leaves. They feed on weeds and grasses, as well as garden plants, and hibernate under trash and in crevices.

Life Cycle: 1. Eggs on leaves 2. Nymph (small adult) 3. Adult

Controls

Biological: No tested biological controls exist, but you can try applying Nc nematodes in the fall to kill overwintering forms. Larval parasites have been identified.

Cultural: Use fall plowing and control weeds.

Acute: Spray soap weekly in the early morning, when insects are less active. Use white rectangular sticky traps to monitor and trap out.

Termites, dry wood *Termitidea family, isoptera order*
Termites, subterranean *Reticulitermes flavipes*

Unlike ants, the winged swarmers and wingless worker termites have a straight thick waist linking their thorax and abdomen. Ants have a narrow "wasp-waist". Termites have straight antennae compared with ants' elbowed antennae. Worker termites are white whereas worker ants are brown or black. Sometimes fly maggots are confused with termites. Maggots have no legs, whereas worker termites have six well defined legs. Both are white in color. As the "subterranean" termite, the most common type, feeds on wood, it eats only between the grain and fills it in with dirt. Subterranean termites require moisture and contact with the soil where their nests are located. "Drywood" termites live above the ground and are rarer in some areas. They cut the wood across the grain and do not fill in with dirt. To inspect for termites, look for signs of a colony both at the soil line and elsewhere (eaves, sills and other wood areas). Drill a hole in the wood if deterioration is suspected, but not obvious. Wood treated with insecticides can be releasing harmful toxins into your home and may present a health risk.

Life Cycle: 1. Eggs 2. Nymph 3. Adult

Controls

Biological: Use beneficial nematodes labeled for termites, such as *Termask*. Apply with injection equipment. Termites seal off members of their colony infected this way, so results may be variable. Soapy water plus nematodes, or soapy water alone, drenched on the colony or nest are alternative controls. Use around boat docks, barns or holiday cabins and your home, if you choose.

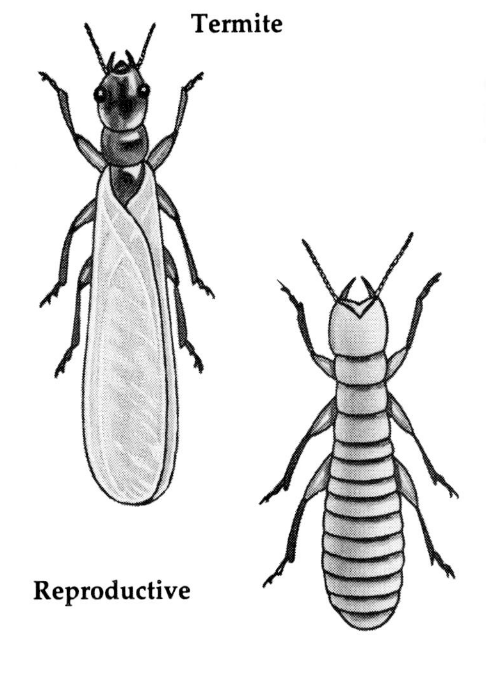

Cultural: When building a home, investigate proper methods of constructing your foundation and foundation-house junction. Brick and concrete foundations help deter termites, but are not totally effective alone. Remove all wood scrap from the building area. Separate all dead wood from the soil. When buying a house, check the design and construction, as well as signs of current infestation.

Acute: Liquid nitrogen treatment injects ultra-cold nitrogen gas into the wood and freezes drywood termites. *Electro-Gun*™ electrocutes termites. Both are only regionally available at this time to commercial pest control agents. Trained dogs can be used to locate colonies. For more information on contacting termite controllers see literature from "IPM Practitioner" and "IPM for Home and Garden". There is also a non-chemical hot air treatment of the entire house for termites and other wood destroying insects. Contact Isothermics, Inc., PO Box 18703, Anaheim CA 92817-8703.

Thrips *Scirtothrips citri (several other species)*

These minute, 1⁄50 inch long, yellow or black insects rasp plants and flowers, sucking their juices and spreading plant disease. When disturbed, they fly, hop short distances and flick their abdomens over their heads like a scorpion. Affected plants drop their blossoms early or have brown leaf tips.

Life Cycle: 1. Eggs on leaves and plants 2. Nymphs (small adult) 3. Adult

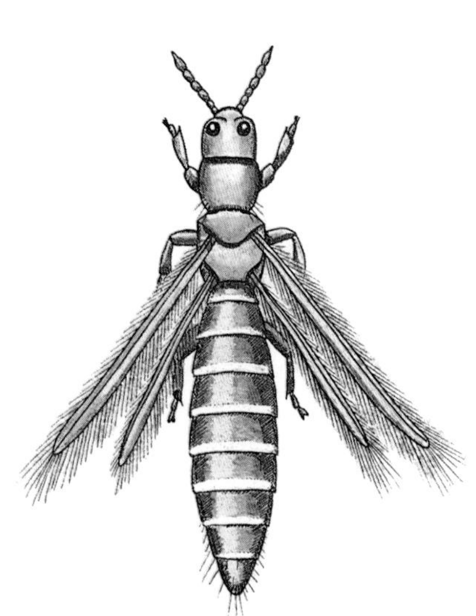

Thrip

Controls

Biological: Two predatory mites are commercially available, *Amblyseiulus mackenseii* and *Euseius tularensis*. Dragonflies prey on thrips. There is one recent report that Nc nematodes in the soil control them in greenhouses.

Cultural: Interplant marigolds.

Acute: Spray soap in the early morning when thrips are least active. Spray a mixture of sugar and sabadilla dust. Use yellow sticky traps.

Tomato hornworms *Manduca quinquemaculata*

These 3 - 4 inch long green caterpillars with white bars feed ravenously on tomatoes, eggplants, peppers, potatoes and some weeds. They have a fleshy "horn, but cannot sting. The adult moth is gray or brown.

Life Cycle: Two generations per year in the South 1. Green-yellow eggs on the lower sides of leaves 2. Larva feed on leaves and fruit 3. Pupate in soil 4. Hummingbird-like moth emerges either in spring in the North or three weeks after pupating in the South

Controls

Biological: Trichogramma, lacewings and ladybug beetles attack eggs. Spray Bt on leaves. Apply Nc nematode as a mulch or top dressing around plants. Handpick unless caterpillars have braconid wasp cocoons in their backs.

Cultural: Fall plow to destroy pupae. Interplant strong smelling herbs.

Acute: Spray Bt. Use lime and soap spray directly on the caterpillars.

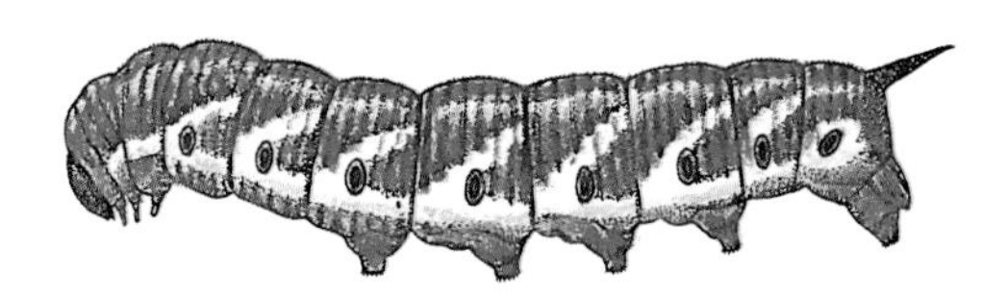

Tomato Hornworm

White grubs

Phyllophaga species

These larval stages of June beetles and Japanese beetles feed on roots. They are comma shaped, up to 2 inches long, and have six legs. It can be found curled in a "c" shape in the soil. It has a well defined head and legs.

Life Cycle: 1. Eggs laid in soil 2. Larva or grub feeds on roots 3. Pupate in soil 4. Adult beetles

Controls

Biological: Apply Nc nematode in spring or early summer as mulch or top dressing. Also see controls for the Japanese beetle. Handpick in soil.

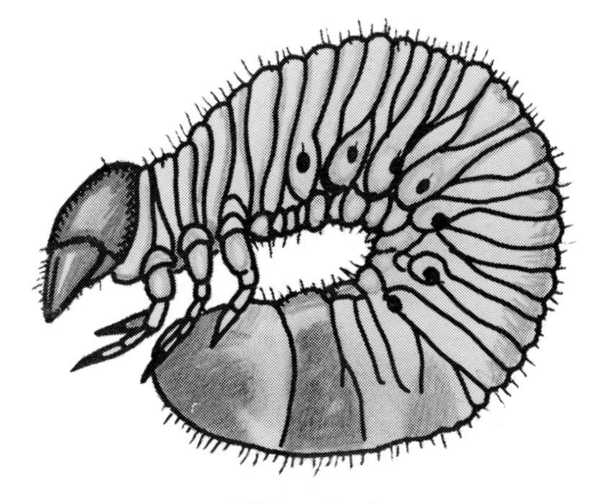

White Grub

Cultural: Use fall plowing. Delay replanting sodded areas for one year.

Acute: Apply milky spore disease, *Bacillus popilliae,* to grassy areas around the garden. It is sold under the trade names *Doom* or *Grub Attack,* as well as others. Use aerator sandels to control all species of white grub larval stage at the right time of the year. Good for orchards and other grassy areas as well.

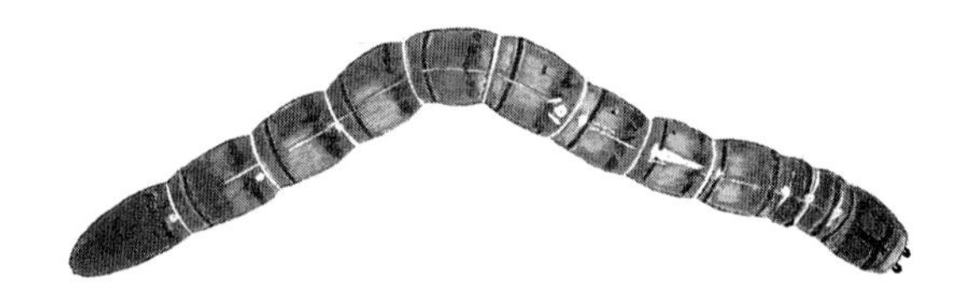

Wireworm

Wireworms *Agriotes lineatus*

These thin worms (larvae) with a leathery skin have three pairs of legs near their heads, whereas millipeds have many more.

Life Cycle: Three to four years 1. Eggs in soil 2. Larva or worm stage feeds on roots 3. Pupate in soil 4. Adult click beetle

Controls

Biological: Apply Nc nematodes two months before planting. Handpick from soil.

Cultural: Plow several times before planting. Delay planting for one year or more on ground that was sod. Plant a fall cover crop. Use fall plowing, crop rotation and leave heavily infested parts of the garden fallow over the spring and summer.

White Flies *Trialeaurodes vaporariorum*

These tannish-white, 1/20 - 1/12 inch winged insects are not real flies, since they have four wings instead of two. They are more like miniature moths with dusty wings. They are particularly troublesome in greenhouses where heavy infestations result in clouds of insects when disturbed. White flies suck juices from leaves.

Life Cycle: 1. Eggs on undersides of leaves 2. Wingless nymph that feeds surrounded by a white, waxy substance similar to mealybugs 3. Pupa 4. Adult

Controls

Biological: The parasite *Encarsia formosa* can be released in greenhouses. Ladybugs and lacewings eat the eggs. Yellow paper or plastic "sticky bars" with tanglefoot trap white flies, but not *Encarsia*. Use the sticky bars alone or with *Encarsia*.

Cultural: Interplant nasturtium and marigolds. Examine all new potted plants brought into the house or greenhouse.

Acute: Spray soap solution or pyrethrin on whiteflies early in the morning. Alcohol added to soap or soap with coconut oil dissolves waxy coating on larvae.

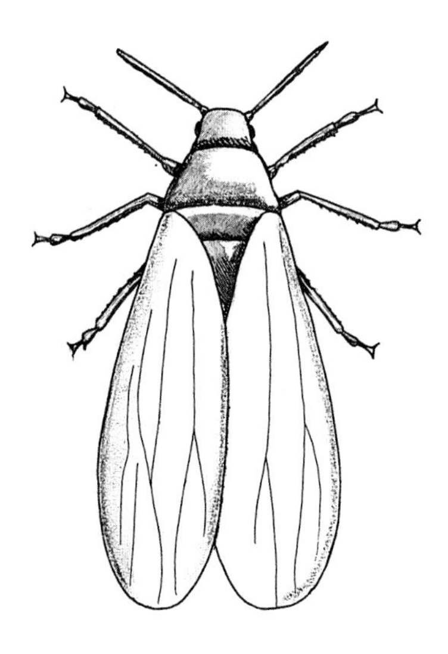

Adult White Fly

APPENDIX

Some Experiments For The Adventurous Gardener

Beneficial nematodes can infect a large range of host insects. You can play a part in increasing what's known about the effect of nematodes on insects in your area by trying the following experiment. Try spraying soapy water with 10,000 nematodes per pint at night or very late in the evening on a cloudy day on the following insects:

Asian Cockroach — Adults
Colorado Potato Beetles — Both adults and larvae
Cucumber Beetles — Adults
Formosan Termite — Adults
Grasshoppers — Adults
Imported Cabbage Worms — Larvae
Japanese Beetles — Adults

Make sure to spray directly on the insect. Use one to three teaspoons of liquid soap per quart. Try one teaspoon first, mixed with the water and nematodes. To determine if the nematodes have infected the insect, wait 24 to 48 hours after spraying and see if any dead insects are lying on the ground where you sprayed the garden. Carefully pick up one or two and place on a folded paper towel or piece of coffee filter. Moisten the paper towel with water, place on a saucer and invert a drinking glass over the dead insect. The glass will keep out airborn micro-organisms that may develop on the insect and obscure the emerging nematodes from view. Keep the paper moist.

The first sign that an infection has started is a change in color of the insect's body, usually to a red, pinkish, or yellow hue. Also, there might be some swelling or puffiness. This will happen from the fourth to sixth day. Keep the saucer out of direct sunlight and add water to the paper towel, if needed.

Around the tenth to fourteenth day chains of nematodes will emerge through the body wall of the insect, usually from the abdomen area. These chains are usually circular in shape. If you have access to a microscope, you can place them on a slide with a drop of water and observe them moving around. You can place the dead insect and the nematodes back in the garden in a well shaded place on moist soil and they will seek out other insects to infect. As many as 200,000 nematodes will emerge from a large grasshopper or cabbage worm.

You might try capturing two or three insects and place them in a jar with a leaf they eat. Use a medicine dropper full of soapy water and nematodes and place two to four drops directly on each insect, or carefully give them a light spray

of the mixture in the jar. Place a lid on the jar and keep away from sunlight. If the insects die in 24 to 48 hours, the nematodes have infected the insect. Proceed with putting the insect on a moist paper towel under a glass as above and observe to confirm your findings.
If you still have trouble getting an infection, place some moist soil mulch with nematodes in the bottom of a jar. Place some cucumber beetles or cabbage worms on top of the mulch and screw on a jar lid with air holes in it. In one or two days the beetles should be dead. These experiments will also give you a chance to try different ways of applying nematodes and test their effectiveness.

Glossary

Bt — *Bacillus thuringiensis,* an insect microbial, affecting most species of caterpillar

Bti — Special strain of Bt that affects only larvae of mosquitos and black flies

Compost — Decaying plant matter that has decomposed enough to be a good source of plant food. Naturally occurs in woods under layers of fallen leaves; a major component of forest soil. Also called humus or leaf mold.

Green manure practices — Also called cover crop. Growing clover, vetch or rye and plowing plants back into the soil just prior to seed bud formation. Object is to build soil fertility.

Herbicide — Group of petroleum based products which inhibit plant growth. Used for weed and grass control. Newer herbicides use species specific plant pathogens.

Hydro-carbon or petroleum based insecticides — Group of chemicals produced from petroleum distillation by-products.

IPM — Integrated Pest Management

Larva — Stage in the life cycle of some insects between an egg and pupa. Generally takes the form of a soft-bodied caterpillar, grub or maggot. Often referred to as the "worm stage".

Lime — Supplies calcium to the soil and plants and neutralizes acid condition of the soil. A soil test is the only way to accurately determine soil acidity.

Microbials — Naturally occurring microscopic organisms that have been discovered in the environment. They can also be produced on a large scale.

Nc Nematode — One of a number of micro-organisms that live in the soil, microscopic in size and worm-like in appearance. Nc nematodes, of which there are several strains, feed and reproduce inside insects. Other types of nematodes attack plants. Nc is not harmful to humans.

NPV — Nuclear polyhedrosis virus, an insect pathogen or disease. A micro-organism or microbial.

Nymph — Stage in the life cycle of some insects between egg and adult. Life cycle of insects with nymph stage is called incomplete metamorphosis because it does not have a pupa or cocoon stage or larval stage.

Organic — The approach to insect control which uses the least dangerous methods and chemicals to control insects. Includes many homemade sprays and remedies.

Piperonyl butoxide — Ingredient added to pyrethrin products to enhance pesticide effect. Be careful with use and handling of piperonyl butoxide.

Pupa — Stage in the life cycle of some insects between larva and adult. Usually inactive inside a cocoon or hard shell found in the soil.

Tanglefoot — A sticky substance made from castor oil, gum resins and vegetable waxes.

Trichogramma — Tiny insect that lays its eggs inside other insect's eggs, causing the host eggs to die.

Some Suppliers of Beneficial Insects and Organic Products by Mail Order

Send for price list before ordering. Commercial quantities are also available. Include a self-addressed stamped envelope. For a more complete list see IPM Practitioner Publications Catalog.

Ringer Corporation
9959 Valley View Road
Minneapolis, MN 55344
(Full line of natural, environmentally-sound lawn and garden products)
(800-654-1047)

Beneficial Insectary
14751 Oak Run Rd
Oak Run, CA 96069
(Beneficial Insects and fly parasites, ladybugs and organic products)
(916-472-3715)

BioLogic
P.O.Box 1
Willow Hill, PA 17271
(717-349-2789)
(Scanmask nematodes and Termask nematodes)

Great Lakes IPM
10220 Church Road
Vestaburg, MI 48891
(pheromones)

Beneficial Insects Inc.
P.O. Box 40634
Memphis, TN 38174
(beneficial insects)

Rincon-Vitova Insectaries
P.O. Box 95
Oak View, CA 93022
(800-248-2847)

Natural Pest Control Co.
8864 Little Creek Dr.
Orange Vale, CA 95662
(mosquito fish)

Necessary Trading Co.
New Castle, VA 24127
(703-864-5103)

Peaceful Valley Farm Supply
PO Box 2209
Grass Valley, CA 95945
(916-272-4769)
FAX (916-272-4794)
($1.00 for catalog)

Richters
Goodwood
Ontario L0C 1A0
CANADA

ARBICO
Arizona Biological
P.O. Box 4247
Tucson, AZ 85738
(800-505-BUGS)

Growing Naturally
PO Box 54
Pineville, PA 18946
(215-598-7025)

NGMMG
RD 1, Box 715
Landisburg, PA 17040
(gypsy moth parasites)

Gerhart Inc.
6346 Avon Belden Rd.
North Ridgeville, OH 44039
(216-327-8056)
(leafminer parasites)

Pest Management Supply
P.O. Box 938
Amherst, MA 01004
(pheromones)

Down To Earth
850 West 2nd
Eugene, OR 97402
(Distributor of full line of organic products)

IPM Laboratories
Locke, NY 13092-0099
(315-497-3129)

Some Manufacturers of Microbials (Micro-Organisms) and Other Products

Attack Pesticides
Division of Ringer Corporation
9959 Valley View Road
Minneapolis, MN 55344
(800-654-1047)
(Manufacturers of a broad range of microbial pest controls)

Abbott Laboratories
Dept. 95 M
1400 Sheridan Rd
N. Chicago, IL 60064
(*Dipel*™ BT and *Vectobac*™ Bti, a strain of *Bacillus Thuringiensis* specifically for mosquitoes. Contact for local distributors.)

BioChem Products /BPG
Box 4090
Kansas City, MO 64101
(*Bactospeine*™ Bacillus Thuringiensis and *Bactimos*™ Bti—inflowable concentrate or wettable powder. Contact for distributors in your area.)

BioLogic
418 Briar Lane
Chambersburg, PA 17201
(717-263-2789)
(*Scanmask*™ Nc nematodes for gardens and *Termask*™ Nc nematodes for
termites)

Sandoz Inc. / ZOECON
12000 Ford Road, Suite 400
Dallas, TX 75234
(*Thuricide*™Bt, *Teknar*™ Bti, *Elcar*™ Nuclear Polyhedrosis Virus, pheromone lures and monitoring traps)

Crop Genetics /Dupont / Monsanto
7170 Standard Drive
Hanover, MD 21076
(301-796-4633)
(NPV for European Pine Sawfly and other NPV's)

Safer Agro-Chem Inc.
(See Ringer Corp
9959 Valley View Road
Minneapolis, MN 55344)
Safer's Insecticidal Soap (cut with 30 % ethyl alcohol and 20% water)

Tanglefoot Co.
314 Straight Ave
Grand Rapids, MI 49504
(Tree Tanglefoot. Contact for distributors.)

JTLK, Inc.
PO Box 427
Boonton, NJ 07005
(manufacturer of head lice comb and non-pyrethrin soap, *Derbac* —excellent non-chemical control)

Bronner's Pure Castile Soap
Box 28
Escondido, CA 92025

Mycogen Corp.
5451 Oberlin Dr.
San Deigo, CA 92121
(manufacturers of Bt-sd)

Ecogen
2005 Cabot Blvd
Langhorne, PA 19047
(215-757-1590)
(*Cutlass*™ Bt)

TADD Services
1617 Old County Rd., Suite 4
Belmont, CA 94002
(termite-seeking dogs, available to pest control contractors nationwide)
(800-345-TADD)

Eco Science
85 North Whitney
Amherst, MA 01002
(413-256-8985)
(Non-genetically altered microbes)

Ladd Research Industries
PO Box 1005
Burlington, VT 05401
(Ladd apple maggot trap)

ASPCI
P.O. Box 7414
Berkeley, CA 94707
(Codling moth virus)

Pesticides and Water Quality

A large volume of pesticides is used in urban areas as well as farming fields and orchards. These chemicals leach into underground water systems, wells and streams where they can travel great distances. With two pounds of insecticide used per year per household, water quality is not just a rural concern, but an urban one as well. Some pesticides can cause birth defects or cancer.

For more information on what you can do to discourage the use of pesticides, contact:

Northwest Coalition for Alternatives to Pesticides
PO Box 1393
Eugene, OR 97440

National Coalition Against the Misuse of Pesticides
530 7th Street
Washington, DC 20003

Citizens for Alternatives to Chemical Contamination
11463 Bringold Avenue
Lake, MI 48632

Write your Congress person:

The Honorable (Your Senator)
United States Senate
Washington, DC 20510

The Honorable (Your Representative)
United States House of Representatives
Washington, DC 20515

For more information on treatment for and identification of pesticides, contact:

Your local Poison Control Center

National Pesticide Hotline
Department of Preventive Medicine
Texas Tech University
Lubbock, TX 79430 (800-858-7378)

Center for Bio-Organic Studies
Room 200
University of New Orleans
Lakefront
New Orleans, LA 70148
(Have a doctor call 504-286-6644)

National Medical Services
Willow Grove, PA 19090
(Have a doctor call 215-657-4900)

Natural Resources Defense Council
Toxic Substances Information Line
Outside New York state (800-648-NRDC)
Inside New York state (212-687-6862)
122 East 42nd Street
45th floor
New York, NY 10168

Enviro-Health Systems
990 North Bowser Road. Suite 800
Richardson, TX 75081
(Have a doctor call 800-558-0069)

For legal assistance you or your lawyer may contact:
The Association of Trial Lawyers of America
Environmental and Toxic Tort
Litigation Section
1050 31st, NW
Washington, DC 20007

Bibliography

Brand, Stewart et. al., *The Next Whole Earth Catalog*, Second Edition, 1981 Point, Random House
British Museum of Gardening History, *Forsythe on Trees*
Carson, Rachel L., *Silent Spring*, 1962, Houghton-Mifflin
Carr, Anna, *Rodale's Color Handbook of Garden Insects*, 1979, Rodale Press
Field Enterprises, *World Book Encyclopedia*, Chicago, 1955, Vol. A (Ants) & Vol. I (Insecticides)
Greystone Press, *New Illustrated Encyclopedia of Gardening*, 225 Park Ave South New York, NY 10003
Hunter, B. H., *Gardening Without Poisons*, 1964 Houghton-Mifflin, Boston
Loudon, J.C., *Encyclopedia of Gardening*, 1824
Matsumura, F., *Toxicology of Insecticides*, 1985 Plenum Press New York, NY
Metcalf, R., et al, *IPM For Home and Garden*, 1980 Univ. of IL I.E.S. Urbana, IL 61801
Milius, S., "Growing Food, Poisoning Water", *Organic Gardening Magazine*, September 1986, Vol. 33, No.9
Milius, S., "If You're Sprayed", *Organic Gardening Magazine*, July 1986, Vol. 33, No. 7 (Buyer's Guide edition lists chemicals and equipment)
Olkowski, William and Helga, and Sheila Daar, *IPM Practitioner*, 1986 P.O. Box 7414, Berkeley, CA 94707
Philbrick, Helen and John, *The Bug Book*, 1974 Garden Way Publishing Pownal, VT
Poinar, Prf. G. O., *Nematodes For Biological Control of Insects*, 1979 CRC Press
Pyenson, *Fundamentals of Entomology and Plant Pathology*, AVI Publishing Westport, CT 06881
Rodale, Robert, *The Basic Book of Organic Gardening*, 1971, Rodale Press
Swan, L. A., *Beneficial Insects*, 1964 Harper and Row, NY
Van Den Bosh, Robert, *The Pesticide Conspiracy*, 1978 Doubleday and Company
Von Strum, Carol, *A Bitter Fog: Herbicides and Human Rights*, Sierra Club Books, 1983
World Health Organization, "Heptachlor: Environmental Health Criteria 138", 1984, WHO Pub. Center, 49 Sheridan Ave., Albany, NY 12210
Yepsen Jr., R. B., *Encyclopedia of Natural Insect and Disease Control*, 1984 Rodale Press
Make Compost in 14 Days, 1982, Rodale Press Editorial Staff, Emmaus, PA, 18049

Addendum to revised edition:

1. "Death Trap — Carnivorous Plants", PBS *Nature* series. Available from Time-Life Videos
2. DeBach, P., ed. *Biological Control of Insects and Weeds*, 1964, Chapman and Hall

continued on next page ...

3. Cook and Baker, *The Nature and Practice of Biological Control of Plant Pathogens* , 1983, American Phytopathological Society
4. Granodos and Federici, *The Biology of Baculoviruses Vol. 1 and 2* , 1986, CRC Press (NPV's and CMGV)
5. Olkowski, Daar, and Olkowski, *Common Sense Pest Control* , 1991, Taunton Press, Newtown, CT, 715 pages (Least Toxic Control: Lawn, Garden and Community)

For Commercial Growers

Of interest to commercial growers and others interested in how commercial crops are grown, including consulting entomologists:

New Farm Magazine
Rodale Publishing
Emmaus, PA 18049

American Vegetable and Greenhouse Growers Magazine
Willoughby, OH 44094
(Buyer's Guide edition lists chemicals and equipment)

IPM Practitioner
PO Box 7414
Berkeley, CA 94707
(pamphlet titled "What is I.P.M.?" and Publications Catalog $1.00)

Central Farm & Family Center
PO Box 3330
Des Moines, IA 50316
(Farm supplies and tractor parts)

Pixall Corp.
Clear Lake, WI 54005
(Green bean harvesters)

Porter-Way Harvester Corp
Waterloo, NY 13165
(Leaf harvesters)

Stanhay Precision Seeders - Asgrow Seed Co.
PO Box 8
Mechanicsburg, PA 17055

Piedmont Plant Company
PO Box 424
Albany, GA 31703
(open field vegetable transplants — free catalog)

Journal of Alternative Agriculture
9200 Edmonston Road
Suite 117
Greenbelt, MD 20070

Young Entomologist Society (Y.E.S.)
Dept of Entomology
Michigan State University
East Lansing, MI 48824

Additional Reading Section

Must reading for all homeowners and gardeners:

"I.P.M. for Home and Garden"
$2.00 Payable to University of Illinois
408 South Goodwin Ave
Urbana, IL 61801

Organic Gardening Magazine
Emmaus, PA 18049

IPM Practitioner (good scientific reading)
$3.75 single issue
Vol.8 No.4, April 1986
PO Box 7414
Berkeley, CA 94707
(latest beneficial insect producers issue, updated every year. See all back issues.)

Common Sense Pest Control Quarterly
PO Box 7414
Berkeley, CA 94707
(good general reading on insect control for home gardeners and home owners)

Beneficial Insect Producers U.S. and Canada

Do not purchase insects except from companies in your own country.

American Sport Fish Hatchery
PO Drawer 20050
Montgomery, AL 36120
(205-281-7703)

Applied Bionomics
11074 W. Saanich Rd
Sydney, BC CANADA V8L 3X9
(604-656-2123) FAX (604-656-3844)

ARBICO
PO Box 4247 CRB
Tucson, AZ 85738
(800-767-2847) or (602-825-9785)
FAX (602-825-2038)

Beneficial Insectary
14751 Oak Run Rd
Oak Run, CA 96069
(800-477-3715) or (916-472-3715)
FAX (916-472-3523)

Better Yield Insects
PO Box 3451 Tecumseh Station
Windsor, Ontario
CANADA N8N 3C4
(519-727-6108) FAX (519-727-5989)

Bio Collect
5841 Crittenden St
Oakland, CA 94601
(510-436-8052)

Biofac, Inc
PO Box 87
Mathis, TX 78368
(512-547-3259) FAX (512-547-9660)

BioLogic
PO Box 1777
Willow Hill, PA 17201
(717-349-2789)

Biological Control of Weeds
1140 Cherry Dr
Bozeman, MT 59715
(406-586-5111)

Biosys
1057 East Meadow Circle
Palo Alto, CA 94303
(800-821-8448) or (415-859-9500)

Biotactics, Inc
22412 Pico St
Grand Terrace, CA 92324
(714-783-2148) FAX (714-681-7915)

BoBiotrol
54 South Bear Creek Dr
Merced, CA 95340
(209-722-4985)

Bozeman Bio-Tech
1612 Gold Ave
PO Box 3146
Bozeman, MT 59722
(406-587-5891) FAX (406-587-0223)

Buena Biosystems
PO Box 4008
Ventura, CA 93007
(805-525-2525) FAX (805-525-6058)

Bunting Biological
PO Box 2430
Oxnard, CA 93034
(805-986-8265)

Canadian Insectaries
5 Alderwood Rd
Winnipeg, Manitoba
CANADA R2J 2K7
(204-257-3775) FAX (204-256-2206)

C.H. Musgrove
2901 Everwood Dr
Riverside, CA 92503
(714-785-1680)

CRS Company
2909 NE Anthony Ln
St. Paul, MN 55119
(612-781-3473)

East Arkansas Fish Distributors
PO Box 361
Hazen, AR 72064
(501-255-3455)

FAR
510½ West Chase Dr
Corona, CA 91720
(714-371-0120) FAX (714-737-0718)

GIBCO BRL
8400 Helgerman Court
Gaithersburg, MD 20877
(800-828-6686) or (301-840-8000)
FAX (800-331-2286)

Harmony Farm Supply
3244 Hwy 116 No. 15
Sebastopol, CA 95472
(707-823-9125)

Hydro-Gardens, Inc
PO Box 9709
Colorado Springs, CO 80932
(800-634-6362) or (719-495-2266)
FAX (719-531-0506)

IPM Laboratories, Inc
Main St
Locke, NY 13092
(315-479-3129)

Inslee Fish Farm, Inc
PO Box 207
Connerville, OK 74836
(405-836-7150)

J. Harold Mitchell Co
305 Agostino Rd
San Gabriel, CA 91776
(817-287-1101)

J. Reilly
5000 Trenton St
Matairie, LA 70003
(504-887-3666)

J.M. Malone and Son Enterprises
PO Box 158
Lonoke, AR 72086
(501-676-2800) FAX (501-676-2910)

Keo Fish Farms
PO Box 123
Keo, AR 72083
(501-842-2872) FAX (501-842-2156)

Kunafin Trichogramma Insectaries
Rte 1 Box 39
Quemado, TX 78877
(800-832-1113) or (512-757-1181)
FAX (512-757-1468)

M & R Durango Inc
PO Box 886
Bayfield, CO 81122
(800-526-4075) or (303-259-3521)
FAX (303-259-3587)

Nat'l Gypsy Moth Management Group Inc
RD 1 Box 715
Landisburg, PA 17040
(717-789-3434)

Natural Pest Controls
8864 Little Creek Dr
Orangevale, CA 95662
(916-726-0855)

Nature's Alternative Insectary
Box 19 Dawson Rd
Nanoose Bay, BC
CANADA V0R 2R0
(604-468-7912)

Nature's Control
PO Box 35
Medford, OR 97501
(503-899-8318) FAX (503-899-9121)

Nematec
PO Box 93
Lafayette, CA 94549
(510-735-8800)

Praxis
PO Box 360
Allegan, MI 49010
(616-673-2793) FAX (616-673-2793)

Richmond Fisheries
8609 Clark Rd
Richmond, IL 60071
(815-675-6545)

Rincon-Vitova Insectaries, Inc
PO Box 95
Oak View, CA 93022
(800-248-2847 outside CA) or
(805-643-5407) FAX (805-643-6267)

Rocky Mountain Insectary
PO Box 152
Pallisade, CO 81526

Sea Ranch
Route 2 Box 604
Sheridan, AZ 72150

Sterling International
15916 E Sprauge Ave
Veradale, WA 99037
(800-666-6766) or (509-926-6766)
FAX (509-928-7313)

The Ladybug Company
8706 Oro-Quincy Hwy
Berry Creek, CA 95916
(916-589-5227)

Whiskers Catfish Farm
216 Thornton Lane
Bowling Green, KY 42104
(502-842-2555)

International Producers

AUSTRALIA

Biocontrol Ltd
PO Box 515
Warwick, Queensland 4370
AUSTRALIA
(076-614488)

Bugs for Bugs
28 Orton St
Mundubbera 4626
AUSTRALIA
(071-654576) FAX (071-654626)

Hawkesbury Integrated Pest Management
PO Box 436
Richmond, NSW 2753
AUSTRALIA
(045-701331) FAX (045-701314)

BELGIUM

Biobest
Ilse Velden, 18
2260 Westerlo
BELGIUM
(32-14-231701) FAX (32-14-231831)

FRANCE

CO. DEA
CREAT del Chambre d'Agriculture
Quartier la Baronne
06610 La Gaude
FRANCE

CTIF L/SAP
Centre de Balandran
30127 Bellegarde
FRANCE

Duclos S.A.
BP3, 13240 Septemes les Vallons
FRANCE
(91-519056) Telex (430 113 F)

ETS Rene Briand
La Gibraye
44430 Saint Sebastien sur Loire
FRANCE

GIE La Croix
Rue du Pont, 21
29213 Plougastel-Daoulas
FRANCE
(98-403030) FAX (98-042437)

Insectarie de Provence
La Bergerie, Les Iles
84840 Lapalud
FRANCE
(90-403082)

SICA-CAF
Rue de Biot
06560 Valbonne
FRANCE
(93-421789)

GERMANY

Conrad Appel
Bismarckstr. 5g
6100 Darmstadt
GERMANY
(06151-852200)

Barmann
Fabricius Str. 2
4010 Hilden
GERMANY
(02103-51233)

Bio Nova
Boschstr. 16
4190 Kleve
GERMANY
(02821-8940)

Gartnerei Hatto Welte
Maurershorn 10
7752 Insel Reichenau
Bodensee
GERMANY
(07534-7190)

KWS
Postfach 146
3352 Einbeck
GERMANY
(05561-311390) FAX (05561-311322)

W. Neudorff
Abt. Nutzorgardsmen
An der Muhle 3
Postfach 1209
D-3254 Emmerthal 1
GERMANY
(49-5155-62460) FAX (49-5155-6010)

Sautter & Stepper
Rosenstr. 19
7403 Ammerbuch 5
Altingen
GERMANY
(070352-75501)
FAX (0149-70327-4199)

GREAT BRITAIN

Biological Pest Control Ltd
Acorn Nurseries, Chapel Lane
West Wittering, Chichester
West Sussex, PO20 8QG
GREAT BRITAIN

Bunting & Sons, The Nurseries
Great Horkesley, Colchester, Essex
GREAT BRITAIN
(020-627-1300) Telex (987385 Bunting G) FAX (020-627-2001)

English Woodlands, Ltd
Hoyle Depot, Graffham Petworth
West Sussex, GU28 0LR
GREAT BRITAIN
(07986 574)

Humber Growers
Common Lane
Welton Brough
North Humberside
GREAT BRITAIN

Natural Pest Control, Ltd
Yapton Road, Barnham
Bognor Regis
West Sussex PO22 OQB Yapton
GREAT BRITAIN
(024-355-3250) FAX (024-355-2879)

HUNGARY

Budai Cs
Hodmesvasarhely Institute of Soil & Crop Protection
Conservation Service
PO Box 99
6801 Hodmezovasarhely
HUNGARY
(46-611) Telex (84-224 noeval)

ISRAEL

Biological Control Insectaries
Kibbutz Sde Eliyahu
D.N. Bet-Shean, 10810
ISRAEL
(06-580527) or (06-580509)

Amos Rubin, Entomologist
(Only Distributor of some species)
1 Keren Hayesod St
Givat Shmuel 51905
ISRAEL
(03-357469) FAX (972-3-5351308)

ITALY

Azienda Agraria Giardina
S.P. 104
Contrada Carrozieri Milocca
90600 Siracusa
ITALY
(0931-721444)

BIOLAB — Centrale Ortifrutticola
Alla Produzione
Via Dismano 3845
1-47020 Pievestina di Cesena (FO)
ITALY
(0547-330434) FAX (0547-318451)

Urbio s.r.l.
via Ghiarino
640056 Crespellano (BO)
ITALY
(051-960244)

THE NETHERLANDS

Brinkman B.V.
(A Distributor of Bunting Inc)
Postbus 2, 2690 AA's-Gravenzande
THE NETHERLANDS
(01748-11333) Telex (32671 BRMA NL) FAX (01748-16517)

De Groene Vlieg
Duivenwaardsedijk 1
3244 Lg Nieuwe Tonge
THE NETHERLANDS
(01875-1862)

Koppert B.V.
Veilingweg 17
2651 BE Berkel en Rodenrijs
THE NETHERLANDS
(31-1891-40444) FAX (01891-15203)

POLAND

Zaklad Produkcji Enteromotagow "Erna"
Nowy Sacz
Wielopole 86
POLAND
(321-78) FAX (0326393)

SCANDANAVIA

Chr. Hansens Bio Systems
Boge Alle 10-12
DK-2970 Horsholm
DENMARK
(011-42-76-66-66)
FAX (011-42-76-54-55)

Kemira Oy
Box 14
02270 Espoo 27
FINLAND

L.O.G.
Okern Torgvei 1
N-0580 Oslo 5
NORWAY

Anticimex AB
c/o Trogardshallen
S-25229 Helsingborg
SWEDEN

Svenka Predator AB
Box 14017
S-250 14 Helsingborg
SWEDEN

Ticab AB
Bokgatan 4
S-232 35 Arlov
SWEDEN
(+46-40-43-76-70)

SWITZERLAND

Nordwestverband
Basel
SWITZERLAND

Versand Ab RGB
Lyonstrasse 10
4053 Basel-Dreispitz
SWITZERLAND
(061-3314040)

International Distributors of Koppert Products

BELGIUM

Aster Devrieze N.V.
Molenstraat 2
2638 Reet
BELGIUM
(03-8884090)

Casteels N.V.
Mechelban 73
2861 O.L Vr. Waver
BELGIUM
(015-755529)

Groenewoud, J. BVBA
Singel 8
2510 Mortsel
BELGIUM
(03-4402001)

Zaadhandel Hollandia N.V.
Westkaai 8
2170 Merksem
BELGIUM
(03-6470607)

CANADA

Safer Ltd
465 Milner Ave
Scarborough, Ont
CANADA M1B 2K2
(416-291-8150)

CHANNEL ISLANDS

Stan Brouard Ltd
PO Box 383, Landes du Marche
Vale, Guemsey
CHANNEL ISLANDS
(0481-52521)

CZECHOSLOVAKIA

EcoTrade
Netolická 11
370 12 Ceské Budejovice
CZECHOSLOVAKIA
(38-40260)

DENMARK

G.G. Garta A/S
Nordholmen 5, Postboks 1130
2650 Hvidovre
DENMARK
(3149-1144)

FINLAND

Oy Schetelig AB
Martinkylantie 52
01720 Vantaa
FINLAND
(0-852061)

TukoGardenia Oy
Kytomaantie 14
04200 Kerava
FINLAND
(0-6191)

FRANCE

Koppert France S.â.r.l.
Lot. Ind. du Puits des Gavottes
147 Avenue des Banquets
84300 Cavaillon
FRANCE
(9078-3013)

GERMANY

G.A.M. van der Goes
Georg Hench Strabe 13-2
8400 Nürnberg 80
GERMANY
(011-328062)

J. Mertens B.V.
Vergelt 3, Postfack 8319
5990 AA Baarlo, Neiderlande
GERMANY
(04707-9292)

R. Babler
Neckarblick 36
7122 Besigheim
GERMANY
(071-4335813)

W. Neudorff GmbH KG
An der Mühle 3
3254 Emmerthal 1
GERMANY
(051-5562460)

GREECE

Harantonis
58500 Skyda
Pellis
GREECE
(381-89802)

HUNGARY

Koppert Hungaria Kft.
Moricz Zs. u. 11.
2500 Esztergom
HUNGARY
(33-11268)

IRELAND

Hortico Industries Ltd.
New Haggard
Lusk, County Dublin
REPUBLIC OF IRELAND
(0001-437620)

ITALY

Koppert Italia S.r.l.
Via Porrettano Nord 37/2
40043 Marzabotto (BO)
ITALY
(051-931356)

THE NETHERLANDS

Koppert B.V.
Veilingweg 17 PO Box 155
2650 AD Berkel in Rodenrijs
THE NETHERLANDS
(31-01891-15203)

NORWAY

Plantevern-Kjemi
Huggenes Gard
1580 Rygge
NORWAY
(9-261177)

POLAND

Rol-Eko Co. Ltd
05-500 Piaseczno
Musiadio, Warszawa
POLAND
(022-568241) or (022-567386)

PORTUGAL

Neoquimica
Av. Defensores de Chaves 35-6
1000 Lisbon
PORTUGAL
(01-532264)

SPAIN

Agrofresas S.A.
Chalet El Pozuelo, Ctra. Rabida - Moguer
Apartado de Correos 71
21800 Moguer
SPAIN
(55-372443)

Complejo Asgrow Semillas S.A.
Zurbano 67-2 B
28010 Madrid
SPAIN
(1-4420399)

Olcosa
La Moura 38
15008 La Coruna
SPAIN
(81-253928)

SWEDEN

Svenska Predator AB
Knut Pais Vag 8, Box 14017
250 14 Helsingborg
SWEDEN
(042-201130)

SWITZERLAND

Leu & Gygax AG
Fellstrasse 1, Postfach 30
5413 Birmenstorf AG
SWITZERLAND
(056-851515)

U.S.A.

Gerhart Inc
(Koppert Distributor)
6346 Avon Belden Rd
N. Ridgeville, OH 44039
(216-327-8056) FAX (216-353-9547)

Plant Science Inc
(Koppert Distributor)
342 Green Valley Rd
Watsonville, CA 95076
(408-728-7771)

Got an Insect Problem

THINK

1. PARASITES

2. PREDATORS

3. PATHOGENS

The BBC Method of Insect Management

Bugs, Beetles and Caterpillars

1. Bugs
    - A. Insecticidal Soap
    - B. Parasites and Predators
    - C. Yellow Sticky Traps
    - D. Pathogens

2. Beetles
    - A. Parasites and Predators, Nematodes
    - B. Pathogens: Fungi, Bacteria, Protozoa
    - C. Funnel Traps
        1. Funnel Design Light Traps
        2. Pheromone Traps
        3. Vacuum Traps
    - D. Botanical Pesticides
        1. Rotenone
        2. Pyrethrins
        3. Sabadilla

3. Caterpillars
    - A. Parasites and Predators —
        Nematodes, Wasps, others
    - B. Pathogens —
        Bt, NPV
    - C. Traps for Moths —
        Pheromone, Light

INDEX

Ask your store to carry our fine line of books or
you may order this book and other fine titles directly from:

THE BOOK PUBLISHING COMPANY
PO Box 99
Summertown, TN 38483

or call: **1-800-695-2241**

A Cooperative Method of Natural Birth Control	$ 6.95
Climate in Crisis:	
The Greenhouse Effect and What We Can Do	$11.95
Ecological Cooking: Recipes to Save the Planet	$10.95
From A Traditional Greek Kitchen	$ 9.95
George Bernard Shaw Vegetarian Cookbook	$ 8.95
Judy Brown's Guide to Natural Foods Cooking	$10.95
Kids Can Cook	$ 9.95
Murrieta Hot Springs Vegetarian Cookbook	$ 9.95
Nature's Chicken	
The Story of Today's Chicken Farm	$ 4.95
The Now & Zen Epicure	
Gourmet Cuisine for the Enlightened Palate	$17.95
The NEW Farm Vegetarian Cookbook	$ 7.95
A Physician's Slimming Guide, Neal D. Barnard, M..D.	$ 5.95
Also by Dr. Barnard:	
The Power of Your Plate	$10.95
Live Longer, Live Better (90 min. cassette)	$ 9.95
Beyond Animal Experiments (90 min. cassette)	$ 9.95
Shepherd's Purse: Organic Pest Control Handbook	$ 9.95
Shopping Guide for Caring Consumers	
A Guide To Products That Are Not	
Tested on Animals	$ 5.95
Spiritual Midwifery (Third Edition)	$16.95
Starting Over: Learning to Cook with Natural Foods	$10.95
The Tempeh Cookbook	$ 9.95
Ten Talents (Vegetarian Cookbook)	$18.95
Tofu Cookery	$14.95
Tofu Quick & Easy	$ 6.95
The TVP Cookbook	$ 6.95
Uprisings: The Whole Grain Bakers' Book	$13.95
Vegetarian Cooking for Diabetics	$10.95

Please add $2.00 per book for shipping.